Jonathan Bernstein

Sector Trading: A Year in Exchange Traded Funds

Maerska
New York, New York

A Maerska Financial Book
Published in the United States by
The Voss Rohde Publishing Group, Inc.

Copyright © 2006 by Jonathan Bernstein

M 10 9 8 7 6 5 4 3 2
ISBN 0-9772945-0-1
First Trade Edition

Certain content from articles published
on ETFzone.com are included here
by arrangement.

All rights reserved. This publication may not be
reproduced, stored in any retrieval system or transmitted
in any form or by any means, electronic, mechanical,
photocopying, recording or otherwise, without the prior
permission in writing from the publisher and author.

On the web at: www.maerska.com
ETFzone on the web at: www.etfzone.com

Typeset in the United States by AkVektor
Printed and bound in the U.S.

Maerska
New York, New York

Good investing is really just common sense. But it is astonishing how few people have common sense.

Jim Rogers

Contents

Why ETFs? A Trader's Story...i

Introduction...1

Part I, Q3 2004

Are REITs and Bonds Rate-Sensitive?..10

What Does the EMA (Exponential Moving Average) Tell Us?...........14

Does the Price of Crude Affect the Technology Sector?...................21

Do Round Numbers Matter?...25

Why Buy Energy ETFs?..28

Is the Trend Your Friend?...32

How and When to Go Against a Trend?.......................................35

When the Market Is Rallying: Is It Too Late to Buy?.......................38

How Does the Price of Crude Impact Energy ETFs?.......................42

Which ETFs Are "Politically" Correct? 45

"Sell in May and Go Away," but Is Fall a Good Time to Buy?...........49

What Does the Transportation Sector Tell Us?..............................52

Oil and Tech Redux: Does the Inverse Relation Maintain?..............55

Is the Bull Back?...57

Part II, Q4 2004

How Important Is the Monthly Employment Report?......................63

Randomness: Is There Always a Reason? 67

What Does the Fed Model Tell Us?..71

Are Utilities a Buy?...76

What Does a Bush Victory Mean for the Market? 81

What Is the Technical Outlook for Fixed-Income ETFs?..................85

With Election Euphoria Over, Where Will the Market go Next?.........89

What Happens on Thanksgiving Week?..93

What Does Counting Tell Us?...95

What Do I Do When the Market Trades Sideways?.........................99

Mergers in the Software Sector – Time to Buy?............................102

The Pfizer Case: Can One Holding Crash an ETF? 104

Are Trends in 2004 Important for 2005?.....................................108

Part III, Q1 2005

What Do We Do When Everything Goes Wrong?...........................113

Defense or Offense?...118

Will Technology in 2005 Be a Repeat of 2004?............................122

What Is an "Untradable Sector?"...128

What Do OPEC's Announcements Tell Us?..132
Why Are Long-term Bonds Outperforming Short-term Bonds?..........138
What Does the PPI (Producer Price Index) Tell Us?............................142
Fear or Greed?...149
Biogen in Trouble: Sell the Sector?...154
How Important Is It to Be Right?..160
What Is the Easy Trade?...164
What Does U.S. Inflation Tell Us About Foreign ETFs?.....................168
What Is a Fakeout?...171

Part IV, Q2 2005

What Happens When Oil Collapses?..177
What Does April 15th (Tax Day) Tell Us?..181
Can Broad Market Volatility Be Used to Predict Oil Prices?.............186
What Does Earnings Season Tell Us?..190
Is the Re-Introduction of the Long Bond Good for Investors?.............194
How to Protect the Ego When Losses Mount?.....................................198
What Made the Biggest Rally of the Year?..202
Nasdaq Up Eight Days Straight: Time to Buy?...................................206
Summer Is Here: Time to Sell Volatility?..210
Why Are Treasury Yields So Low?...215
Is Alternative Energy a Viable Oil Play?...221
What Are Fundamental Drivers of Oil and Utilities?...........................225
In Conclusion..231

Appendices

Glossary 240
List of ETFs...243
ETF Index..247
Index..249

Why ETFs? A Trader's Story

Starting the Millennium on the Wrong Foot

I knew if I lost any more money, I'd end up one of the ghosts haunting the trading floor bringing coffee and Chinese food to senior traders. All the ghosts wanted one thing: to convince the management to let them back in the game. It really depended on how much money you lost, but the basic rule was that you could have your terminal back and a small stake if you managed to recruit ten new guys for the company. That was pretty easy in 1999 and 2000 – a little bit of cold-calling, posting up signs at NYU or ponying up some personal cash to put an ad for the company up on Jobtrak.

But this was January 2001, the new millennium, and things were different. The Nasdaq was crashing, and suddenly no one wanted to be a daytrader anymore. The ghosts were all over the floor – guys on H1B visas from France, Germany, Korea, even Africa, who had to sign back their mandatory paychecks to the company. The ghosts weren't fools. They were smart guys with MBAs who walked around talking about "regression lines," "oscillation stochastics," and other financial mumbo-jumbo that nobody understood, practicing their pitch on anyone who would listen. Really all they wanted was a single-screen terminal and trading privileges. For now at least, I still had a terminal and a log-on, but they had changed my seat three times in the past two months – a sure sign that my period of grace was coming to an end.

One of the mythical stories about traders is that they need the right psychological environment to be successful. Trading, it is thought, is partly a question of identity. Management tries to group people together, like with like, to facilitate success. The trading gurus and therapists are part Deepak Chopra, part Transcendentalist, like Thoreau and Ralph Waldo Emerson: "Know yourself, be yourself," they advise. The chorus is loud: If you are yourself when facing the market, you will be successful; you will make money.

For some reason management started me out in 1999 with a bunch of rowdy hippy dudes: dreadlocked white guys who smoked a lot of dope, took their profits when the summer doldrums hit, went to Thailand's Ko-Samui and rented the best damn beach condo money could buy. Then they lined up the dope and the girls. But during the trading day, these guys wouldn't even watch CNBC. They unplugged cable, stuck a VCR onto the wall monitor and popped in porn videos. Noise at the desk was the usual combination of triumph and agony, all the while mixed in with the sounds of humping from the TV set. Everybody had different market positions, so while one guy would be screaming "Oh yeah!" another guy would be so upset that he'd be breaking down, banging his fists on the table, tearing his mouse apart, throwing it at the monitor, all the while cursing his position in the market, and his lot in life – that is until someone from MIS came over and got him a new mouse.

When I didn't make money trading with the hippy dudes, management put me with a team of West Point grads, mostly black guys, who did listen to CNBC during the day and were much more cool-headed than the hippy set. They were deep buddies, dressed slickly, had identical top-of-the-line 911 Porsches. They drove down to Wall Street every day in a convoy from somewhere in Connecticut, chatting on their cell phones. I'd get there early and could see them occasionally out the office window cruising into the garage on Water Street at about 9 a.m., 30 minutes before market open. They had a buddy who worked over at Paine Webber who fed them information about big customers. Their game was to front-run the PW customers for 10,000 or 20,000 shares. That didn't really hurt the PW customers too awful much but it did help these guys to make a living. They were real nice, the desk was quiet, and we got along well, but sitting with them I was supposed to contribute by coming up with a deal similar to the PW one. I didn't have any friends in the business or on the floor, and so eventually they didn't share their information with me any more.

Management figured it wasn't working out and moved me across the street to a trader named David's group – mostly orthodox Jews. A mezuzah – a box with a piece of scroll from the Torah – hung on the door to the trading floor. The guys were famous at the firm for doing really well. A lot of them wore yarmulkes and black suits and had strings hanging from their belts. They had beards, hats, wore all-black clothes. When the market was closed you could usually find them hanging out together talking earnestly, often in Hebrew, which I don't understand. But once in a while, I'd catch one of them alone in the elevator and ask: "How yah doing? Shalom." The answer was inevitably formal, religious: "Thanks be to God." Figuring at first that they didn't understand my question, I said, "I mean in the market – you up money today? You find a trade today, man? Were you in Kodak at the open?" "God is glorious" would be the response. "God is glorious."

It was getting pretty obvious in the early days of 2001 that my P&L (profit and loss ratio) was not improving. With this group, as with the other groups before it, I traded deeper and deeper into the hole. This is a difficult place to be, and anyone who has ever traded knows what it feels like to get up, go to work, work all day, and yet return home poorer than you left in the morning. A down day becomes a down week, which becomes a down month. It is very tough.

So slowly I put together my own rather desperate plan to make back the money I owed the company. The plan went like this: Wait for GLW Corning – an important fiber company whose stock had run up a lot in 1999 and 2000 – to open below 50 and then all at once break my DOT limit (my trading limit set by the risk management area of the company) and rapidly send multiple market sell orders down to the floor and get very very very short GLW to save my ass from becoming one of the ghosts.[1]

I'd been planning for two weeks, when on February 7, the day finally came. GLW opened up at $49.10 and went lower. I sold short, held to the close, and the risk-management team looked the other way about my breaking my DOT limit (as they predictably did on winning positions). I did the same thing for the next three days on 10,000 shares each day, and by the end of the week, I had just about wiped out my debt to the company. When I got back to zero, I quit daytrading, quit Wall Street and went back to California to think it over.

Lessons Learned

Lessons learned as a daytrader have informed my subsequent trading. Three key realizations provide the foundation for this book and the viability of sector trading using ETFs.

The first realization is that trading individual stocks is more difficult than trading sectors. A focus on individual companies is problematic because the internals of any single organization's business are difficult for any outsider (and even most insiders) to understand. In a year, Enron went from the seventh-largest company in the world, capitalized at near $100 billion, to bankruptcy. Tyco. WorldCom, Adelphia Communications – insiders, or anyone with access to specific information has an edge over the ordinary investor who tries to make decisions based on tape reading, price movement, or even fundamentals.

Rather than think in terms of an individual company's stock, I

---

1   Short-selling is betting that a stock will go down in price. It involves an investor selling something he does not own. Typically, the investor borrows shares from someone on the street with the promise to return them at a later date. The trader then sells the shares in the market. If the stock falls, he is able to buy them back and return the borrowed shares, keeping the difference between his sell and buy price as profit. If instead a stock goes up, he will be forced to pay more for a stock than he has sold it for, and the position will become a loss.

started to think in more macro terms, in terms of *sectors* that offered exposure to several companies with products in that area: for example, the financial sector, or the oil sector, or the semiconductors. Why sectors? Because as a market observer without insider knowledge, if I have an edge, then it lies not with stock-picking or micro-analysis of any individual company or issue. It lies not with tape-reading or analysis of the micro-movements of a single stock. If I have an edge, it lies in catching larger sector-wide trends, rather than, say, whether or not news of the Debian 3.1 enterprise system from management services company RedHat (RHAT) will work out better than investors expect, and therefore cause the stock to rise. As markets – not just in the United States but worldwide – increasingly trade in lockstep, this kind of stock-picking is becoming more difficult to do well.

The second realization is that most traders overtrade. When I started trading, I was in and out several times a day, sometimes short and long on the same stock within an hour. Reviewing my trades I realized that I almost never made money when I traded from both the short and the long side of the same stock during the same day. As daytraders we had learned to be nimble, get in and out quickly (and this racked up a lot of money in commissions). But my second realization was that I needed to keep my trades in for a longer period.

The third realization is that many traders, and particularly daytraders, spend too much time thinking about specific entry and exit points and worrying about the spread. In a typical trade for example, I would look for large blocks of shares printing on a big downtick at the figure, and then try to buy there, while everyone was selling and the NYSE specialist (I hoped) getting long. I had become, to quote Jesse Livermore, an "average ticker hound... [gone] wrong from over-specialization."[2] I needed to find a way to broaden my thinking and simplify the mechanics of trading so that I could focus on longer-term trends and larger issues where I might have an edge. ETFs provided the entry point into this equation.

---

2   Lefevre, Edwin, *Reminiscences of a Stock Operator*, p. 60. New York: John Wiley and Sons, 1993

I'd like in particular to thank Will McClatchy for his inspiration, Paul Weil and Vicky Elliott for their swift and expert reading. And a very special thanks also to the dear Theodore Storgion, Jason Draho, and Rosalee Shim.

I hope that the book proves as helpful for readers as it has for me.

Jonathan Bernstein
San Francisco, August 2005

1

Introduction

Books on investing tend to be either heavily quantitative, with a textbook-style approach to details, or chirpy and anecdotal – amusing to read but ultimately more entertaining than practical. This book is neither. It is a handbook that provides a weekly survey of a year in the markets from the perspective of Exchange Traded Funds (ETFs).

Since they began trading in 1993, ETFs have arguably become the most important new product for the retail trader and investor. ETFs are appealing because they offer high tax efficiency and low expense ratios, and can be traded intra-day. In the last five years, ETFs have begun trading for every major sector in the market – from technology to oil to real estate. There are also bond ETFs as well as ETFs representing the markets in every major country in Western Europe, Latin America, and Asia. ETFs are like mutual funds in the sense that they are a composite of holdings in individual companies. But unlike mutual funds, ETFs are trading vehicles. They trade all day long, whenever markets are open. There are currently no actively managed ETFs. Technically, ETFs are certificates that state a legal right of ownership of the many individual stock certificates which make up the holdings.

Before the advent of ETFs, in order to make a sector trade, an investor had to either buy or sell a basket of stocks one stock at a time, which involved significant transaction costs, or buy from a mutual fund company, which is often inflexible and limited. Though

~1~

some traders did pioneer swing trading in mutual funds, because the costs of this trading are borne by all the fund holders, companies soon moved to discourage the practice by creating penalties for exiting before a pre-specified length of time.[1] These rules make mutual funds poor vehicles for trading. ETFs, by contrast are usually highly liquid, with tight bid-ask spreads. Typically ETFs do not have the turnover of mutual funds and therefore do not distribute the capital gains mutual funds do. As a result most ETFs can be held like stocks for many years with few tax consequences. ETFs also have superior tracking compared to CEFs (Closed End Funds), which often trade at wide discounts or premiums. In addition, scandals now have tarnished the mutual fund industry, as companies have allowed big traders and favorite clients to profit at the expense of investors. The ETF market remains scandal-free.

For many years, traditional funds had an advantage over ETFs simply because there were no ETFs to cover many of the asset classes important to investors. This is no longer true. Since my first ETF trade in 2000, when there were only a handful of ETFs available to trade, the number and variety of ETFs has increased dramatically. There are now ETFs for virtually every sector in the market including bond ETFs, ETFs specialized in real estate, health care, technology, and even in gold bullion. There are ETFs for virtually every sovereign country in Western Europe, for Latin America, all the important Asian economies and recently two Chinese ETFs. Over three quarters of sector specific ETFs have put and call options available.

These features make sector ETFs premiere vehicles for trading. One of the best ways for an individual investor who has an opinion on the macro environment and is able to identify trends is to trade ETFs. Trading an ETF is as easy as trading an individual stock. It requires a funded brokerage account and a computer with an Internet connection. Nothing more. It has never been so easy for the average investor to have an opinion about a trend in the market and

---

1   One of the early mutual fund timers was Gil Blake, who took advantage of a Fidelity policy that at the time allowed unlimited switching of funds between accounts at no cost. Fidelity no longer offers such funds. While trading ETFs does involve transaction fees, depending on the brokerage, these fees can be very low. Unlike many mutual funds, there are no minimum holding periods for ETFs and no penalties for frequent transactions.

to seek to profit by placing a bet in one of the many markets represented by ETFs worldwide.

Trading With ETFs

Let me give an example of a kind of trade that ETFs have made available to the retail investor: after the Madrid bombing in March 2004, and again after the Spanish election where the voters put the socialist government in power, the Spanish market was badly hurt. News of the bombing, the election, and the market sell-off made the front page of almost every newspaper in the United States. An investor reading the news might believe that the sell-off in Spain, for example was not material to the trend of higher prices for Spanish securities and want to invest in the Spanish market. This opinion could be articulated by buying an ETF that tracks the Spanish market. The Spanish ETF is called the iShares MSCI Spain Index (EWP).

Before the introduction of this ETF, a retail investor might have placed a bet on the Spanish market by buying an individual component of that market traded as an ADR on the New York Stock Exchange – Telefonica, S.A. (TEF) for example. This would not be a bad bet to make, given that Telefonica is an important stock for Spain. It is a huge company that provides fixed-line voice telephone, wireless communication, and Internet services. Telefonica can be expected to correlate reasonably well with the broad Spanish market. But there are problems, too. Telefonica has significant investment in Latin America, for example, which an investor perhaps does not expect to be influenced by the bombing in Madrid or the elections there. Not only that, Telefonica might react differently for any number of other reasons particular to its specific business area. It might have poor earnings, or any other number of disappointments that single companies – even a large and established companies – are prone to have. In other words, it is not clear how good a proxy Telefonica is for the general Spanish market. As a result, the trader risks being right about the Spanish market but wrong about Telefonica. However, trading the EWP, the Spanish ETF, which seeks to provide investment results that correspond to the Spanish market overall, broad exposure is possible. Buying the EWP, a trader don't have to worry about the stock mix. The ETF itself already uses a sampling strategy to track Spain.

ETFs are also good vehicles for developing hedged, or short-long positions. Unlike individual stocks, most ETFs can be sold short on a downtick. And by buying and selling put and call options, investors can add additional leverage to a portfolio or create positions that move on specific sector or broad market volatility. Increasingly sophisticated bond strategies involving the shape of the yield curve and even credit spread trades are possible using fixed-income ETFs in combination.

This book is primarily introductory and focuses mainly on more basic uses of ETFs. The book seeks to familiarize readers with key ETFs and how they actually trade on a day-to-day basis. It will demonstrate how to use the movement of ETFs as a tool for understanding, evaluating, and profiting from broad market trends. The book introduces a mentality as well as a practical trading methodology. It will be of value to individual investors as well as to analysts who seek to articulate broad perspectives on market direction, focus a portfolio, or provide speedy and flexible exposure to a sector. For readers unfamiliar with specific ETF terminology and language, a basic glossary is included as an appendix. There is also a summary description of all ETFs mentioned in the book.

The Question Method

The specific format of the book is a week-by-week analysis of the market as it appears from the perspective of ETFs. Consisting of fifty-two chapters representing fifty-two weeks in the market, each chapter reviews the price action of key ETFs during that week and poses a question for the reader to consider. This design is a variant of the case study method, a favorite for teaching MBA students to think quickly and strategically. Though this method is not often used in the investment world, it should be; investors are dependent almost entirely on written reports for decision making and have to weigh a variety of factors.

Experience with the market, unfortunately, does not come cheap. Watching a portfolio stagnate or decline while markets take off is the way most investors learn. This book attempts to help both new and seasoned investors to short-cut this painful process by considering various situations that arise in the markets. In the course of looking at this material the reader will become familiar with ETFs as an asset class as well as a variety of theoretical models for interpreting price action.

The Portfolio

In order to follow the course of these situations, a portfolio is developed as of the first week and modified each week, consisting of picks made based upon fundamental and technical criteria. The portfolio format is inspired by George Soros' "real-time experiment" in the *Alchemy of Finance*,[2] and includes analysis derived from articles written over the course of the year for ETFZone.com and Yahoo! Finance. For the sake of simplicity, the size of every position held in the portfolio is expressed in generic ETF unit or units. All portfolio changes therefore are understood to be hypothetical and pedagogical and, to avoid overtrading and overtiming, considered to be initiated only once a week (Monday morning) for orders submitted pre-market, receiving in every case the opening print Monday on the American Stock Exchange or other relevant exchange. For existing positions, the portfolio is analyzed and calculated as of the close Friday. In cases when ETFs were dropped from the portfolio, they were exited the following Monday at the open. Portfolio implications are slightly, but not materially, different.

Over the course of the year, the portfolio came to consist of the ETFs listed in the table below:

---

2   Soros, George, *Alchemy of Finance,* New York: John Wiley and Sons, 1987. As in Soros' book, commissions and other specific transactional data are omitted. Of course, there is no question that commissions are important and that they can be the ruin of any trader who trades frequently and does not heed them.

ETFs Traded in the Portfolio	
AGG	iShares Lehman Aggregate Bond Fund
DIA	Diamonds Trust
EWZ	iShares MSCI Brazil (Free) Index Fund
IBB	iShares Nasdaq Biotechnology Index Fund
ICF	iShares Cohen & Steers Realty Majors Index Fund
IDU	iShares Dow Jones U.S. Utilities Sector
IEF	iShares Lehman 7-10 Year Treasury Bond Fund
IGM	iShares Goldman Sachs Technology Index Fund
IGW	iShares Goldman Sachs Semiconductor Index Fund
PBW	PowerShares WilderHill Clean Energy Portfolio
QQQQ	Nasdaq-100 Trust
SPY	Standard and Poor's Depositary Receipts (SPDRs)
TLT	iShares Lehman 20+ Year Treasury Bond Fund
XLE	The Energy Select Sector SPDR
XLF	Financial Select Sector Index Fund

Table 1: Portfolio Description

Weekly changes to the portfolio will be noted at the conclusion of each chapter in the format of the following table:

ETF	LW	TW	P/L
XLE	5	4	↓1.9%
IDU	1	2	↑2.6%
SPY	1	1	↑0.3%
QQQQ	0	-1	
Portfolio +1.2%			

LW: Last Week | TW: This Week | P/L: Profit/Loss

Table 2 : Example Portfolio Table

This table above has four columns. The first column on the left shows the name of the ETF. The second column (LW) shows Last Week's position size. The third column (TW) shows This Week's position size. In the fourth column (P/L), the Profit and Loss of each position from Last Week (LW) to This Week (TW) is represented as a percentage.

In the case of the portfolio above, the size of the Energy Select Sector SPDR (XLE) was "5," or 500 shares. The position was cut back to "4," or 400 shares. The decision to cut back the position by

~Introduction~

100 shares, leaving 400 shares, was made based on the 1.9% loss suffered between Last Week (LW) and This Week (TW). In the second row of the first column, IDU or iShares Dow Jones U.S. Utilities, is listed. IDU gained 2.6% from Last Week (LW) to This Week (TW). Based on these gains an additional 100 shares were added to the position, doubling it in size to "2," or 200 shares. In the third row of the table above, the 100 share position of the Standard and Poor's Depositary Receipts (SPY) is up slightly. But the table shows that there is no change in position size: LW and TW are alike. In the fourth row, the Nasdaq 100 Trust (QQQQ) shows a 0 in the Last Week (LW) field. This means that there was no position in the prior week, and so as yet there is no Profit or Loss. The table in fact indicated that this position is being initiated. Also, under the This Week (TW), field the QQQQ shows -1. The negative sign means that as of This Week (TW), a *short* position is being initiated. Finally, the bottom row shows the overall portfolio's gain or loss for the week. The portfolio gain or loss calculation factors in the real dollar cost of the positions. For example, although XLE is five times the size of the SPY in terms of the number of shares, on a dollar basis the two positions are actually similar. The reason for this is that the share price of XLE in the fall 2004 at $30-$35/share is about one third of the SPY, priced at over $100/share. So in the above example, even though XLE lost 1.9% on 500 shares, it may be inferred from the positive performance of the portfolio overall that gains in the higher price per share SPY and IDU more than offset the losses in XLE. Reading the table may seem complicated at first, but after reviewing a few case studies, the reader will quickly become familiar with the format.

⚠ *Tips or unique market movement appear in italics with this symbol preceding.*

Although pedagogically I think the book is best understood by beginning with Week 1 and reading through to Week 52, investors interested in seasonal trends may find it useful to skip around from chapter to chapter, and to use the book as a reference tool for consulting price, sector and market behavior on specific dates. Because the market is notoriously prone to seasonality, topics organized by date and covered in the chapters below may provide indications of future price behavior.

To make this demonstration simple and to provide a tool for readers, there are 10 main indicators that influence the market and that no investor can afford to miss. They are:

Market Influence Indicator Descriptions	
GD	Government Economic Data (everything from employment to trade numbers)
SF	Seasonal Factors
CC	Currency, Global Issues
EN	Energy
BY	Bond Yields
IP	Investor Psychology
FA	Fundamental Analysis
TA	Technical Analysis
PD	Political Decisions
MR	Market Rhythm and counting

Table 3: Market Influence Indicators

Market indicators will be displayed at the top of each chapter to provide the reader with a quick means of scanning the material. Of course, there is often substantial overlap between many of these indicators.

Sector Trading: A Year in ETFs traces the history of the markets beginning with the last week of the second quarter in 2004 and ending with the end of the second quarter of 2005. While this particular one-year period is chosen somewhat arbitrarily, the end of second quarter of 2004 is a good beginning point for a number of reasons. First, on June 30, 2004, the last day of the second quarter, the Federal Reserve raised rates for the first time in four years, marking a watershed moment. Second, the beginning of the second quarter is about half-way to the November election of 2004, which had an important effect on ETFs and on the broad market. Third, the end of the second quarter is the beginning of summer, traditionally the lowest in volume and slowest time of year. Beginning with the summer months provides readers with the opportunity to familiarize themselves with the galaxy of ETFs before the storm of fall.

Part I, Q3 2004

~

Week 1 - Week 14

Week 1: June 28 – July 2

QUESTION:
- *Are REITs and Bonds Rate-Sensitive?*

INDICATORS:
- GD, SF, BY

SITUATION:

It is midsummer, and volume is low. But this doesn't mean that nothing is happening. On Tuesday, June 30, 2004, the Federal Reserve raised rates 25 basis points, the first hike in four years.[1] Then Friday, July 2, the Labor Department reported that 112,000 jobs were created, less than half the 250,000 expected. The chart below shows how bond ETFs traded this news.

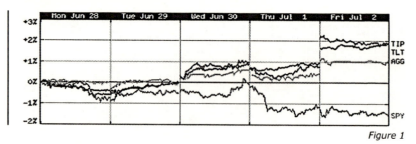

Figure 1

---

1 A basis point is 1/100 of a percentage point. So 25 basis points is 0.0025.

The chart above compares the returns of three bond ETFs with a benchmark equity index fund, iShares Lehman 20+ Year Treasury Bond (TLT), iShares Lehman TIPS Bond (TIP), and iShares Lehman Aggregate Bond (AGG) to the benchmark Standard and Poor's Depositary Receipts (SPY). As the chart above shows, despite the rate hike, bonds did not sell off, in fact ending the week higher.

Friday's jump reflects the weak payrolls number, announced before market open. Bond investors are guessing that a weak economy will encourage the Fed to keep rates low.

REITs (Real Estate Investment Trusts) also overcame the interest rate hikes Tuesday. On their way to a negative week, REITs closed slightly to the plus side. The chart below shows how closely matched are the returns of three REIT ETFs: iShares Dow Jones US Real Estate (IYR), iShares Cohen & Steers Realty Majors (ICF), and streetTRACKS Wilshire REIT Fund (RWR).

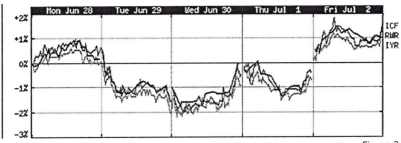

Figure 2

DISCUSSION:

Why should bonds and real estate funds, known as REITs (Real Estate Investment Trusts), be rate-sensitive? Why should bonds or REITs be hurt by rising rates?

Fixed-income (or bond) ETFs are portfolios of bonds. The value of a bond typically decreases as the interest rate set by the Federal Reserve increases. The reason for this is that bonds pay their owners a fixed percentage. Suppose, for example, an investor's portfolio consists of just one newly issued $1,000 U.S. Treasury bond paying 4%. Because the U.S. government issues new Treasury bonds every week, if the Federal Reserve raises rates, this means that when the government issues a new bond, it will have to issue it at a higher interest rate than before the Federal Reserve raised rates. A newly issued bond, sold for $1,000 and with a $1,000 face value, would therefore be paying a higher rate, say, 5%. After the new bond is issued, an investor would rather own a $1,000 bond paying 5% than a $1,000 bond paying 4%. Because the newly issued bond has just been sold for $1,000, the value of the the older $1,000 bond in the investor's portfolio paying 4% must be worth less than the $1,000 bond paying 5%, so it is worth less than $1,000. Of course, an investor still expects to receive the $1,000 and the 4% until the bond's maturity, but if the owner of the 4% bond in this example were to sell the bond before the surprise rate increase, he would receive more for the bond than after. Therefore, when rates go up, it usually costs bond owners money.

REITs are vulnerable to rising rates because they borrow money to fund real-estate purchases, so their interest expenses go up if rates rise. On the other hand, the sentiment that the economy is faltering actually often helps REITs. The reason is that when the economy struggles, cheaper money, in the form of low interest rates, is necessary to maintain growth. With oil prices rising, the health of the economy is threatened, so the Fed is under some pressure to keep rates low. Presumably, Greenspan will continue to raise rates. But the speed and severity of future interest rate hikes will be critical because there are reasons to buy defensive sectors like bonds and REITs: fear of terrorism, stock fatigue, and a weak economy overall.

Investors made the weak economy their main concern, buying bonds for safety on the employment news.[2] So the most important news this week may be not about something that happened, but rather something that failed to happen: Rates rose but the rate-sensitive bonds and REITs failed to go lower. The market has not seen a rate hike in four years, and yet this did not trouble the progress of bonds or REITs. Yes, it was clear that this hike was coming. Yes, investors had four years to prepare for this move. Yes, the move was already priced into the market. But when the hikes actually start, bonds and REITs might still have gone down or closed flat on the week. Instead, they moved higher.

The reason why bonds and REITs held, even appreciating, as shown in Figure 2 above, is because fear of a declining market is

---

2   When equity markets suffer, many investors seek refuge in the relative safety of bonds.

trumping the news of higher interest rates.

With oil prices going up and and fewer jobs being created than expected, fear of an economic slowdown is a lesser evil than a modest rate increase.

PORTFOLIO:

Because the typically rate-sensitive bonds and REITs did not go up on the rate hike in the way they were expected to go up, investors may just now be warming up to the idea that bonds and REITs may not be as sensitive as previously thought. This is a trend that I think will continue into future weeks. I want to position for bonds and REITs to go up again next week. The portfolio will start off long bonds (AGG), long REITs (ICF).

ETF	LW	TW	P/L
AGG	0	1	0
ICF	0	1	0
Portfolio: --			

LW: Last Week | TW: This Week | P/L: Profit/Loss

Week 2: July 6 – July 9

QUESTION:
- *What Does the EMA (Exponential Moving Average) Tell Us?*

INDICATORS:
- TA, IP

SITUATION:
Nobody knows what to do. This market is all over the place, but generally tilted to the downside, and selling. A quick pull of five key sector ETFs demonstrates the confusion.

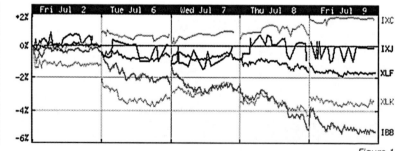

Figure 1

The chart above compares the returns of an oil service fund, iShares S&P Global Energy Sector (IXC), with representative ETFs in the finance, health care, technology and biotechnology sectors. Of the ETFs shown in the chart, iShares Nasdaq Biotechnology

~14~

(IBB) was the worst performer, down almost 5% from last week. The Financial Select Sector SPDR (XLF) and the Technology Select Sector SPDR (XLK) also lost, while iShares S&P Global Healthcare Sector (IXJ) remained unchanged.

Higher oil prices are not helping the market, but they are helping oil companies, the energy sector, and energy ETFs. The question for now is: How much higher will oil prices go? On Friday at the New York Mercantile Exchange, crude closed the session just under $40 a barrel, at $39.96.

High oil, low tech. While oil booms, the technology sector is falling apart. The week's worst performer in tech was the software index (IGV), down 6.4% from last Friday's close. Software was hit by disappointing profits and a poor earnings outlook. The chart below compares IGV with networking (IGN) and the semiconductors (IGW), plus two diversified technology ETFs: iShares Goldman Sachs Technology Index (IGM), and iShares Dow Jones US Technology (IYW).

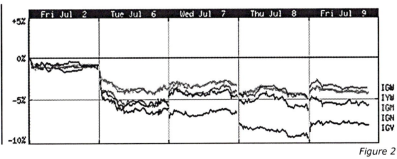

Figure 2

Technically, the ETF that tracks the Nasdaq: the Nasdaq 100 Trust (QQQ), the ETF that tracks the Dow Jones Industrial Average: DIAMONDS Trust (DIA), and the ETF that tracks the S&P 500 Index, the Standard and Poor's Depositary Receipts (SPY), are all sitting on or near their 200-day EMA (exponential moving average). This technical situation is important because both short- and long-term traders watch the 200-day EMA technical indicator to assess price direction.

DISCUSSION:

When building a portfolio, it can be helpful to have a longer-term technical picture, especially when the market is looking around for direction. From a technical perspective, what does the market look like on a longer-term basis?

The two charts below show the QQQ plotted against its 200-day EMA, and the DIA compared to its 200-day EMA. The smoother gray line shows the EMA, which represents the historical price. The darker line shows the price performance of the DIA.

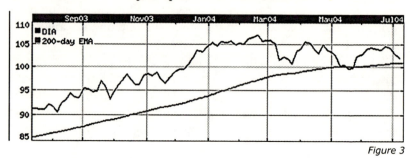

Figure 3

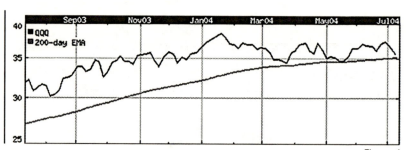

Figure 4

What is the EMA? The EMA is an average price for a security recalculated daily and going back a specific period of time. The moving average is an indicator of how much higher or lower current prices are compared with a historical average price. Because of the relative infrequency of intersection points with the 200-day EMA (as compared to other shorter-term averages such as the 14-day or the 50-day EMA), the 200-day EMA is often watched to provide buy and sell signals.

As the charts in Figure 3 and Figure 4 above show, the broad market DIA and QQQ are trading close to the 200-day EMA. In this situation I watch to see whether the ETF will "bounce off support," meaning trade near or hit the EMA, and then go higher, or whether the fund will "trade through support," meaning that it will hit the EMA, fail to bounce, and go lower. In other words, if an ETF goes below the 200-day, the expectation for many will be that it will continue lower. On the other hand, if the ETF moves significantly above the EMA, many will expect it to continue in that direction.

▲ *If an investor believes that an ETF is going higher in the long run, hitting support by touching the 200-day EMA from above is a signal to buy. One way to trade this is to put a limit buy order just above the point at which the ETF meets the 200-day EMA.* If the ETF does not go higher as expected and trades through support, a trader might sell the long position for a small loss, and wait again for the ETF to push above the 200-day EMA. An ETF breaking the support of the 200-day EMA might be an opportunity to sell the ETF short (and perhaps buy that short position back when the ETF touches the 200-day EMA again).

But because the DIA and QQQ are sitting right on the EMA (neither above nor below) looking at the chart does not tell us whether they will bounce off the EMA or trade through support.

One way to assess the possible direction of broad index ETFs, such as the QQQ, SPY, and DIA, is to look at the performance of the components of these indexes.

▲ *By reviewing where sector-specific ETFs are in relation to their own 200-day EMA, it may be possible to assess overall market direction.*

When also trading very close to their 200-day EMA, sector ETFs provide little insight about the direction of the broad market. For example, the graph below shows the Financial Select Sector SPDR (XLF). Like the QQQ, SPY and DIA, it is resting right at its 200-day EMA.

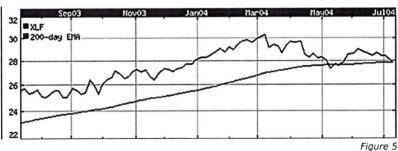

Figure 5

Similarly, a health care sector ETF, the iShares Dow Jones US Healthcare (IYH), is sitting on its 200-day EMA. The chart below shows this:

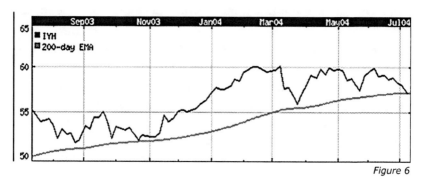

Figure 6

But things look different for the technology sector. If an investor believes that broad market ETFs will follow the lead of technology, as they often have in the last five years, it is notable that most technology sector ETFs have fallen below their 200-day EMA. The chart below shows the 1-year performance of the broad-based tech sector benchmark iShares Goldman Sachs Technology Index (IGM).

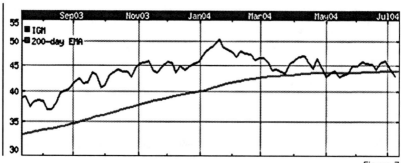

Figure 7

As the above chart shows, IGM has just traded below its 200-day EMA. The 200-day has flattened out since the beginning of May, when IGM first traded below.

Even more convincing is the chart of the biotech sector, iShares Nasdaq Biotechnology (IBB).

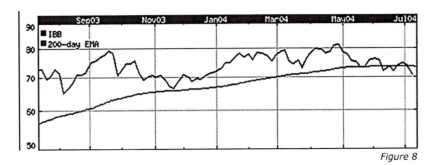

Figure 8

Like the technology sector ETFs, biotech is trading below its 200-day EMA. This is a clear sell signal.

On the other hand, oil and natural resource ETFs are trading significantly above their 200-day EMA. The graph below shows the 1-year performance to date of the iShares Dow Jones US Energy (IYE) and its 200-day EMA.

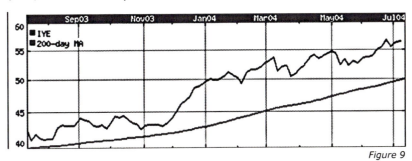

Figure 9

Although the chart is less pretty, REIT ETFs have also moved convincingly above their 200-day EMA. This chart shows iShares Cohen & Steers Realty Majors (ICF).

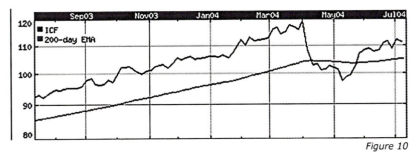

Figure 10

The sum of this technical analysis is that it looks as if technology and biotech are pulling the main indexes lower, and oil and REITs are pulling the main indexes higher. The mix of technology trending lower and oil trending higher leaves it unclear from a technical perspective whether broad-market ETFs like the DIA and SPY (which have both energy and technology components) will bounce off support or trade through it. Technically, technology and especially biotech look awful. This looks like a great opportunity to sell short IGW, IGM, or IBB. Oil might be a good buy, as the trend is up. It might be good to open up a position in the most liquid oil ETF Energy Select Sector SPDR (XLE). But oil is a difficult sector to trade, and the price of oil is mostly dependent on supply, which is very political.

PORTFOLIO:

I'll wait for some more confirmation before buying oil. But I'll start a biotech (IBB) and technology (IGM) short position at these levels.

ETF	LW	TW	P/L
AGG	1	1	↓0.3%
ICF	1	1	↓0.9%
IGM	0	-1	0
IBB	0	-1	0
Portfolio -0.6%			

LW: Last Week | TW: This Week | P/L: Profit/Loss

Week 3: July 12 – July 16

QUESTION:
- *Does the Price of Crude Affect the Technology Sector?*

INDICATORS:
- EN, TA, GD, BY

SITUATION:
For this week, one chart says it all:

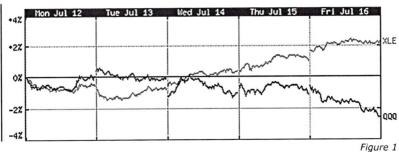

Figure 1

XLE is an oil sector ETF. The QQQ is a proxy for the technology-heavy Nasdaq. The chart shows that as energy ETFs like XLE moved higher, technology fell. The tentative hypothesis is that money is leaving technology and moving into energy. The chart above is beautifully mirror-like. The next chart is less pretty:

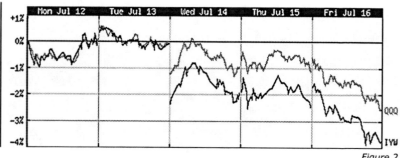

Figure 2

The chart above shows the utter collapse of technology: iShares Dow Jones US Technology (IYW) ended the week down 3.6%. The Nasdaq Composite, which on Friday traded below the statistically important 1,900 number for the first time since May, came within a just few points of its lowest close since 2003.

Separately, Friday was a big day for bonds, which rose on the news that the CPI (Consumer Price Index) was up just 0.3% in June, half of May's CPI number. The chart compares the 1-week return of TLT, AGG, and LQD.

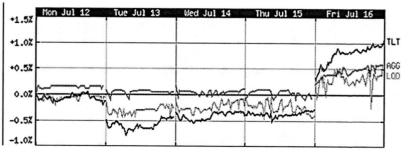

Figure 3

The low CPI number suggests that inflation will be mild. Any sign of mild inflation is good news for bondholders, because this means that the Fed will not be in a hurry to raise rates to prevent inflation. As the chart above shows, among fixed-income ETFs, the long bond TLT was the biggest beneficiary, shooting up more than a percentage point on the news, and closing out the week with a gain of 1.1%. The aggregate bond AGG and corporate bond LQD were up about 0.6% on the week.

DISCUSSION:

Good news for energy ETFs has come with bad news for technology issues. Obviously, technology does not uniquely use oil in production. And clearly, oil is less important as an input for tech companies than for other companies, such as chemicals or pharmaceuticals.

As Figure 2 above shows, bonds moved higher on the week. Higher prices for bonds are often an indication that investors believe that economic growth is stalling. In this case, the high cost of oil, a critical component of so many industries, is threatening to the broad economy.

Oil and technology reflect opposite investment philosophies. Tech represents growth, oil is defensive. Right now, with the price of crude rising, energy is a leading sector, and investment is rotating out of tech. When oil goes up or the economy stumbles, the technology sector drops.

PORTFOLIO:

Oil closed the week on its highs. Technology closed the week on its lows. I'll add to the short IGM and IBB positions with the expectation that the sell-off will continue next week.

▲ *It is rarely good to buy something that ends the week on its low. When a security ends the week on its low, the presumption is that if the week had another day, the stock would likely take that day to drop lower. The contrary is also true. When a security ends a week on its high, this is often a good time to buy.*

Looking at the chart in Figure 3 above, by the end of the week TLT had outperformed LQD and AGG. But the price of AGG held up all week, whereas TLT was more volatile, down early in the week and coming up only at week's end. I don't feel comfortable with the long duration of the bonds in TLT.[1] I think if the Fed

---

1 Duration is a measure of the price sensitivity of a bond to interest-rate changes. Duration describes the weighted average maturity of a bond. TLT is a "long bond" fund because it holds Treasury bonds that do not come due for more than 20 years. IEF is an intermediate-term Treasury bond fund because it holds bonds that come due in 7 to 10 years. SHY is a near-term

continues to raise rates, bonds will be hurt, and particularly funds holding bonds with longer duration, like TLT. So AGG stays in the portfolio for now. But I am ready to sell this position on any sign that rates will go up quickly.

On the subject of REITs, it is a little risky holding ICF going into the June housing starts number next week. But the rate hikes are not yet significant enough to hurt REITs. The REIT position is improving. I'll add to ICF.

ETF	LW	TW	P/L
AGG	1	1	↑0.7%
ICF	1	2	↑1.7%
IGM	-1	-2	↑3.6%
IBB	-1	-2	↑2.2%
Portfolio +1.7%			

LW: Last Week | TW: This Week | P/L: Profit/Loss

Treasury bond fund holding maturities of 1 to 3 years. The longer-term bonds held in TLT are more sensitive to interest-rate changes.

Week 4: July 19 – July 23

QUESTION:
■ *Do Round Numbers Matter?*

INDICATORS:
■ SF, TA, MR, IP

SITUATION:
Federal Reserve Chairman Alan Greenspan cheered investors Tuesday with his remarks that a future rise in interest rates was "likely to be measured." Initially, the market rallied. The chart below compares the week's return of the SPY with a popular tech fund, the iShares Dow Jones US Technology IYW.

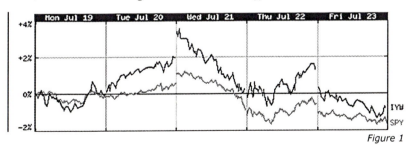

Figure 1

As the chart shows, enthusiasm for Greenspan's remarks held up through Wednesday morning. But then the market headed lower, ending the week on a low and taking the market to new 2004 lows

to key broad market ETFs.

The DIA, designed to mimic the Dow Jones Industrial Average, closed Friday at 99.82, below the psychologically important 100 number, and below 10,000 on the Dow Jones Industrial Average for the first time since May and within a point of its yearly low. The Fidelity Commonwealth Trust (ONEQ), which like the QQQ corresponds closely with the Nasdaq, closed the week at 73.84, its lowest level this year. The chart below compares the 6-month return of the ONEQ and the DIA.

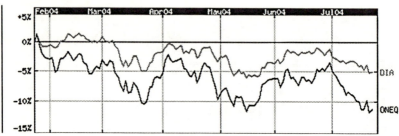

Figure 2

DISCUSSION:

As Figure 2 above shows, the DIA closed below the psychologically important 100. Investors tend to use round numbers as benchmarks. This is true any time a major index ends in "00." With the Dow Jones at 10,000 (and the DIA at 100), this is a very important level, which could easily represent a floor for the broad markets to bounce off and head higher, or a ceiling, which holds the broad indexes down.

Because the DIA has pierced 100 from above, and because the market is on the verge of a yearly low and closed the week near its low, I'm looking at this round number as a ceiling rather than a floor. I think the DIA is in a position to break dramatically lower.

PORTFOLIO:

I'll add a short DIA position and cover it (buy it back) if it closes above 100 next Friday. I also want to sell QQQ on this move to its yearly low.

The REIT position is down 2.6% this week, so I'm getting out of that here, taking a loss before things get worse. I continue to be nervous about holding REITs in a rising rate environment, so the position is counter- intuitive for me. Of course, any trade in summer is tricky. Volume is down. Trading is slow. Trends evaporate.

The tech and biotech shorts continue to work, so I'll add to them here.

ETF	LW	TW	P/L
AGG	1	1	↓0.2%
ICF	2	Closed	↓2.6%
IGM	-2	-3	↑1.7%
IBB	-2	-3	↑4.1%
DIA	0	-1	0
QQQ	0	-1	0
Portfolio +0.2%			

LW: Last Week | TW: This Week | P/L: Profit/Loss

Week 5: July 26 – July 30

QUESTION:
- *Why Buy Energy ETFs?*

INDICATORS:
- GD, EN, FA

SITUATION:
Economic news this week was mixed. On Monday, investors were looking for 6.65 million on June existing home sales. They got 6.95 million. That looked good. Then on Tuesday, July consumer confidence came out at 106.1, more than four points higher than the market was expecting. That looked good. But on Wednesday, June durable goods orders disappointed, coming in lower than expected. And on Friday, second-quarter advanced GDP slowed to a dreary 3 percent.

Some or all of these economic numbers may have influenced the market, but taken together, what they indicate about market direction is ambiguous. What is unambiguous is the price of crude, which closed Friday at $43.80 a barrel, the highest close in over 20 years. How does the cost of a barrel of oil relate to the cost of a gallon of gas? There are 44 gallons in a barrel of oil, so at $44 a barrel, oil costs about $1 a gallon. Refining costs vary, but the handy rule is that a 44-gallon barrel of oil makes 42 gallons of gasoline.

Higher oil is lousy for most of the market, but good news for the energy sector. After disappointing early in the week, oil ETFs recovered nicely and moved convincingly higher. The best performer was iShares S&P Global Energy Sector (IXC), which closed the week at an even $66 per share – up a solid 2.4 percent and close to an all-time high. The chart below compares IXC with Energy Select Sector SPDR (XLE), iShares Goldman Sachs Natural Resources (IGE), and iShares Dow Jones US Energy (IYE). All four energy ETFs in the chart below are highly correlated.

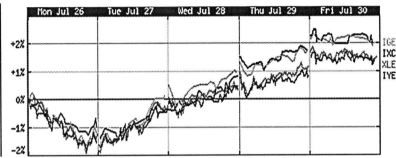

Figure 1

Meanwhile after three weeks of decline, technology ETFs finally had a strongly positive week. The chart below compares the returns of software (IGV) with the semiconductors (IGW), and the more broad-based technology fund (IGM).

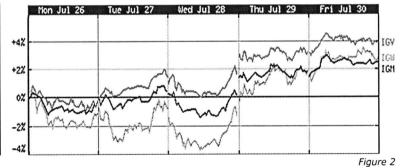

Figure 2

As the chart above shows, leading the pack was the beaten-down software (IGV), up 4 percent. The highly volatile semiconductors are lagging. They were down almost 4 percent mid week before rebounding to close positive.

DISCUSSION:

Economic news is mixed, and crude is hitting record highs. Without any other sector to provide market leadership, the most fertile ground for strength may be in the energy sector. The White House is full of oil elite – President Bush, Vice President Dick Cheney, even the high profile National Security Advisor "I-have-the-ear-of-the-President" Condoleezza Rice.

Even after high taxes, gasoline can easily become more expensive. Consumers may gripe, but most can afford to pay more. So there is plenty of margin here for the oil service stocks and oil ETFs.

Because oil is bought and sold in the futures market, it is not easy for the average investor to make bets on crude oil directly. Oil service ETFs offer comparatively more steady exposure to oil prices and eliminate some of the risk associated with choosing a specific company for investment.

Although crude has been going up and oil companies have looked good for months, with the recent and confirmed weakness in technology, oil looks as if it could be the new leadership sector. Of course, some people have been saying this for months. It is either time to buy oil now in anticipation of new highs or to wait for oil ETFs to take out existing levels and buy when oil goes to new highs. This sector is ending the week on its highs. I anticipate buying when the sector goes to new all-time highs, as it looks poised to do. So I want to be long XLE in the advance of these highs. I want to add it to the portfolio this week.

PORTFOLIO:

This was not a good week for the portfolio. Everything went wrong. Technology advanced, and I am short. The DIA traded back over 100. When the market moves strongly against me, I cut down the portfolio. There is no need to complicate the exit process. It is already complicated enough emotionally, as the ego is involved. Most of the time, the reason to exit a trade is simple: The position is losing money. I shorted the DIA when it traded below 100, a level that I thought could serve as a ceiling. When it traded right back up above that level, the only decision must be to get out. Similarly, I am cutting back the short technology and biotech positions. Why not get out completely? I sold when these ETFs crossed the 200-day EMA, which even after this week's rise, they still have not pierced. Rather than begin covering here, a less conservative trader would be patient and perhaps even short more as the original conditions for the trade obtain. Just as tech weakness was a buying opportunity in the late 1990s, tech strength looks like an opportunity to sell in this environment. But I am always nervous about "doubling down," adding to a losing position. This is even more of a concern in the case of a short position.

▲ *The natural upward drift of the market eventually makes every short position a loser. A good rule of thumb is that any short position should be no more than one-quarter of the size of the portfolio.*

This does not apply universally. Technically, as outlined in Chapter 2 above, there are good reasons for being short right now, but shorting stock long-term is almost never a good strategy. This is especially true when trading sectors, as even the collapse of a single stock within the sector, Enron-style, does not mean that the sector itself will wipe out. In this case, I want to be able to be confident enough to hold on to my current short position. If I short more and technology goes up again, it may be necessary to cover (buy back the short) to prevent losses. A final reason to not get shorter here is that these technology ETFs ended at a weekly high, indicating that buyers may return next week.

ETF	LW	TW	P/L
AGG	1	1	0%
IGM	-3	-1	↓2.8%
IBB	-3	-1	↓3.0%
DIA	-1	Closed	↓1.7%
QQQ	-1	Closed	↓1.9%
XLE	0	1	
Portfolio -2.2%			

LW: Last Week | TW: This Week | P/L: Profit/Loss

Week 6: August 2 – August 6

QUESTION:
■ *Is the Trend Your Friend?*

INDICATORS:
■ GD, TA, BY, IP

SITUATION:
The biggest and most important number for gauging the strength of the economy is the monthly employment report. Investors expected that the biggest economic news this week would be the July employment report. It was. The market had expected that the U.S. economy would add 235,000 nonfarm payroll jobs in July. The Labor Department stunned investors with its report that in fact only 32,000 were added. Huge difference. This news sent most domestic equity ETFs tumbling to new lows for the year.

Bond holders liked the weak employment figure, both because bonds provide security for investors when the economy is weak and because a weak economy usually means that the Fed will think twice about raising interest rates out of fear of starting a recession. Bond prices jumped on the news, closing the week to the plus side. The chart below shows the reaction of long bond TLT, the iShares Lehman TIPS Bond (TIP), which holds bonds that have a provision that protects them from inflation, and corporate bond fund LQD.

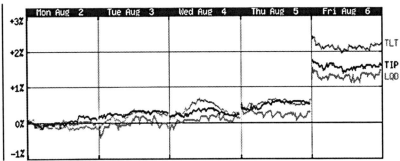

Figure 1

Technology and biotech collapsed. Semiconductors (IGW) suffered the most, down 3.6% on Friday alone and 6.2% on the week. Software (IGV) was not far behind, down 2.4% on Friday and 5.5% on the week. The biotechnology sector ETF fared even worse, down 3.1% Friday and 7.7% on the week. The chart below compares the returns of these three ETFs.

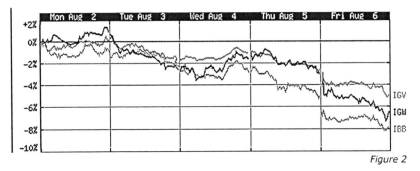

Figure 2

IGM fell right along with the Nasdaq composite, crashing 5%:

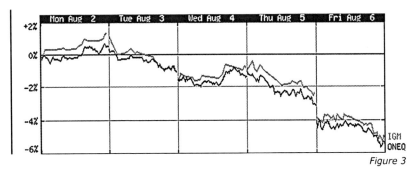

Figure 3

DISCUSSION:

I like to follow trends, and it is rarely a good idea to buy a chart that looks as bad as the chart in Figure 3 above. This looks in fact like a good time to sell. But Figure 3 is not the chart of a lone biotech company or a speculative internet play. It is not even a concentrated sector chart. ONEQ is a proxy for the entire Nasdaq marketplace. The Nasdaq is down 6% in five days. This represents a *huge* move. In addition, this is summertime, so on small volume, the market sometimes makes a big move. I expect a bounce-back.

Sometimes, a good time to take profits is after an unexpectedly big move. The sell-off in technology happened faster and farther than I expected. Also, next week's economic numbers involve inflation, and the markets may use good inflation numbers as an excuse to rally.

A *When there is no news or ambiguous news and the market is attenuated, it often uses seemingly insignificant new economic information as an excuse to reverse, in this case to rally.*

PORTFOLIO:

Big gains on the short side and a big loss on oil. I want to exit all the short positions that benefited from the technology collapse. What to do about the oil position? There are lots of reasons to be in oil. Bush is an oil man, and consumers can afford to pay more at the pump. These are fundamental arguments, and all may be true, but a sharp and immediate loss is a good indication that I don't have it figured out, so I'm out of the oil position right away for a loss.

ETF	LW	TW	P/L
AGG	1	1	↑1.0%
XLE	1	Closed	↓4.8%
IGM	-1	Closed	↑5.9%
IBB	-1	Closed	↑8.6%
Portfolio +3.2%			

LW: Last Week | TW: This Week | P/L: Profit/Loss

Week 7: August 9 – August 13

QUESTION:
■ *How and When to Go Against a Trend?*

INDICATORS:
■ SF, GD, TA, FA, IP

SITUATION:
This was the eighth consecutive negative-to-flat week for the market. Two negatives made this a particularly bad week: oil at $46.03 a barrel, another all-time high, and the June U.S. trade deficit at $55.82 billion, over 20% higher than expected. The best news of the week was a rate hike! The Fed raised rates just 25 basis points. The market moved up in response Tuesday, but drifted lower later in the week, as investors began to worry about high energy prices.

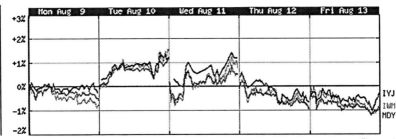

Figure 1

The chart above shows the results of the Fed action. It shows a large-cap fund, iShares Dow Jones US Industrial (IYJ), the MidCap SPDRs (MDY), and iShares Russell 2000 Index (IWM), a small-cap issue.

Especially hard hit was the technology sector: Networking (IGN), semiconductors (IGW), and the more general technology (IGM) sold. The chart below shows this week's sell-off in technology.

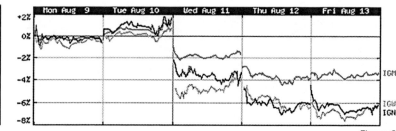

Figure 2

Tech ETFs fell after bellwether Cisco Systems (CSCO) announced disappointing earnings and forecast slower growth pre-market Wednesday. Tech ETFs closed Friday at levels not seen since the fall of 2003.

Bonds held steady, selling slightly on Tuesday's rate hike but moving higher Thursday as the broad market fell. Bonds jumped pre-market Friday on fears of a weaker economy set off by the news that Core PPI rose more than expected. The chart below compares the returns of TLT, IEF, and LQD.

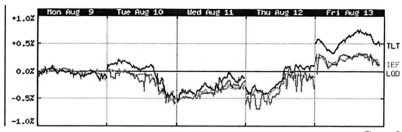

Figure 3

DISCUSSION:

After selling at 6% to 8% per week for two consecutive weeks, I think there is opportunity here to do something I usually hate to do: go against the trend. Why go against the prevailing opinion this time? First, I thought the bounce might be last week and exited all short positions, thereby missing out on this sell-off. There was no bounce this week – things just got worse. Second, this is summer, volume is thin and every big move is fake. Third, a market-wide sell-off of 15% in two weeks definitely invites a contrarian approach.

Of course, there is a huge amount of risk in any contrarian position. When going against a prevailing trend, greed can be fatal. A trader should think: small-size and short-term. The play for a bounce here is a buying a small position in a beaten-down sector here and sell it back at the end of a week or at the end of two weeks max, regardless of any gain or loss. Right now the most troubled sectors are in technology. The trend is still lower. The bears remain in control. But, for a week, the sun may also rise.

In addition to lower tech, as in past weeks the other trend is oil. Crude is up $6 in six weeks, yet the oil service sector has barely budged. This situation looks unsustainable. Why? Historically, oil service stocks follow the price of crude higher. Oil and gas companies, oil drillers, refiners and suppliers, and the integrated energy companies like Exxon Mobile (XOM), ConocoPhillips (COP), and Chevron (CVX) – the kind of companies held in XLE – own oil fields. They touch every part of the process, from exploration to transportation to the sale of the refined product. And oil just got more expensive. Unless crude falls back, these companies will benefit from this. This is a trend to follow: I want to try again with a long position in the oil service ETF XLE.

PORTFOLIO:

Hold bonds (AGG). Add oil (XLE). Short technology (IGM).

ETF	LW	TW	P/L
AGG	1	1	↑0.3%
XLE	0	1	
IGM	0		-1
Portfolio +0.3%			

LW: Last Week | TW: This Week | P/L: Profit/Loss

Week 8: August 16 – August 20

QUESTION:
- *When the Market Is Rallying: Is It Too Late to Buy?*

INDICATORS:
- TA, MR, EN, IP

SITUATION:
Things look rosy. Technology did well, finally. As the chart below shows, the tech-heavy QQQ outperformed the blue-chip DIA, which tracks the Dow Jones Industrial Average and the S&P 500 Index (IVV).

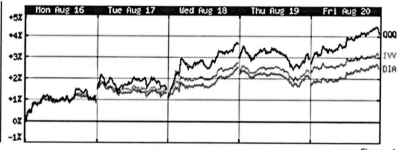

Figure 1

Among technology ETFs, the higher beta outperformed the lower beta issues. Up 10%, the networking ETF IGN had a phenomenal week.

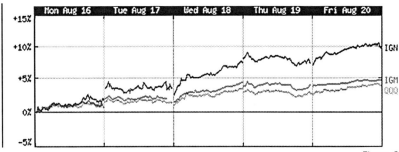

Figure 2

Oil ramped up, coming within a few cents of $50 a barrel, trading to $49.40 intraday Friday, on the news that exports to China (now the biggest consumer of oil in Asia) surged over 40% in July. The energy sector ended the week solidly to the upside. The chart below compares the returns of three energy ETFs, IGE, IYE, and XLE.

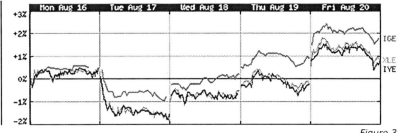

Figure 3

As the chart above shows, IGE outperformed this week, but all three are similar.

DISCUSSION:

Things look rosy. But what is the context of this rally? Is this a "relief rally?" The price of crude is up 20% in seven weeks and at fresh new highs again this week. More importantly, the market is up sharply this week after dropping sharply last week and the week before. And the ETFs rallying this week are the same as those that sold off most sharply last week and the week before.

Before buying in this situation an investor might consider what Victor Niederhoffer calls Lobagola, named after an African bushman who tells of a herd of elephants tramping loudly and dramatically in one direction, tearing up the brush and everything in their path, only to return by the same path a few days later.[1] According to the story, the bushman prepares a trap to catch the elephants in case they return by the same path. Niederhoffer brings the Labagola story to the investment community by comparing the movement of the elephants to the movement of the stock market. During any rally or steep market move, it may help to ask along with Niederhoffer: Can this be compared to the elephants in the African bush?

It is always very difficult to call a new trend, and in this case, the new trend would a positive market going forward for technology. The trend has been to higher oil and lower technology. What has changed? What market conditions would merit such a call? Oil is higher. Technology, though up this week, remains lower than it was two weeks ago.

Rather than suggest a fundamental change in the direction of the market, the tech rally here looks like the return of the elephants by the same path as the sell-off in the last two weeks. The sharpness of the rally, like the sharpness of the drop before it, rather than suggesting a change, reaffirms the pre-rally prices and the bear-market temperament.

In fact, tech looks ripe to short again. The obvious short is IGN, which after heading lower for two weeks, bounced up 10%. This seems like a good short, but IGN may also be returning, Lobagola-style, to where it was. Perhaps the better short is the less visible biotech sector IBB. IBB is a high beta stock and just as sensitive to the market as IGN. And yet, even in this very green week, with moves like the 10% surge in IGN, IBB failed to rally. This is a very bad sign for IBB.

▲ *If the market posts a very strong rally, high beta ETFs can be expected to move up. When a high beta ETF fails to rally under these conditions, this suggests that it is weak.*

IGN may go up again if the market climbs, but I definitely do not want to be long IBB here. With IBB failing to rally on a good week, when everything in the market is higher, what will happen on a bad week, when everything is selling? Will IBB be spared that? Probably not. It looks like a really very good time to go short biotech.

---

1 Niederhoffer, Victor, *Education of a Speculator*. New York: John Wiley & Sons, 1997, pp. 393-394.

PORTFOLIO:

When the positive IGM position was initiated, the promise was to exit that long position within two weeks. The response has been immediate. The position needs to be exited right away.

The fact that crude is moving higher week after week to new records, while oil ETFs trade sideways, is unnatural and suggests that either oil will come down or energy ETFs will pop higher. Two reasons I think oil ETFs will go higher:

1) Oil at a new high may trigger technical buying in oil futures, pulling crude up higher still, and ETFs will follow.

2) The tech and oil relationship: The general inverse relationship between oil and the tech sector remains intact. Because technology looks overbought, when tech sells, money goes into oil. Oil has already moved higher, and energy ETFs could follow. Add to the oil position (XLE).

ETF	LW	TW	P/L
AGG	1	1	0%
XLE	1	2	↑0.9%
IGM	1	Closed	↑4.7%
IBB	0	-1	
Portfolio +1.2%			

LW: Last Week | TW: This Week | P/L: Profit/Loss

Week 9: August 23 – August 27

QUESTION:
■ *How Does the Price of Crude Impact Energy ETFs?*

INDICATORS:
■ EN, PD, GD

SITUATION:
The mood on Wall Street is upbeat. After four weeks of record highs, oil collapsed this week from over $49 a barrel to $43 in just five trading days. Lower oil is big news, and a good reason for general celebration. But there is a puzzle: The shrinking oil price did not hurt oil sector ETFs. Though they fell lower early in the week, by week's end, energy funds XLE, IYE and IGE closed higher. The chart below shows three key oil ETFs: XLE, IYE and IGE.

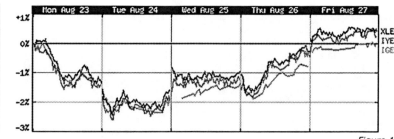

Figure 1

Meanwhile, despite the boost that lower oil is supposed to bring

~42~

to the economy, bonds continue to be strong. All things being equal, this is not consistent with lower oil. If oil falls, the economy should improve and this will cause the Fed to raise rates. Higher rates in turn would cause bond prices to fall. But bonds are up.

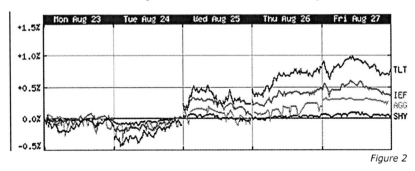

Figure 2

The chart above shows the performance of five bond ETFs: TLT, IEF, AGG, and SHY.

DISCUSSION:

Of course, another way to look at it is that crude is falling because the economy is poor and demand is falling. Seen this way, lower crude prices are a sign of a slackening economy. But I think we are seeing a bull market for oil and investors chasing oil momentum.

The difficult question of course is how meaningful is the current sell-off in crude? There are indications that the current selling is temporary and that in a few weeks, the price of crude will resume a trend higher, and energy ETFs will follow. The first is that the buoyancy of bond prices suggests that the economy is expected to be weak. The second is that there is no broad market celebration, which one might expect if the price of crude were to definitively trend lower. The third is that the trend is to higher oil, and no investor wants to call a trend change.

Last week's analysis was right: Either oil prices fall to oil ETF levels, or oil ETFs rise to oil price levels. But it was also wrong. The expectation was that oil ETFs would rise to match high crude prices. Instead, the reverse happened: Crude fell to oil ETF levels.

This week's results suggest that crude is more volatile than oil ETFs. Also, the movement of crude oil and the movement of oil ETFs are poorly synchronized over the near term. Oil fell $6 (12%) in five days. Energy ETFs closed the week unchanged. When the oil price and oil ETFs move in opposite directions, this always looks like an opportunity. But it is unclear whether I should exit the trade with the expectation that oil ETFs will follow crude lower, or get longer oil ETFs with the expectation that crude prices will rise higher again.

In the short term, oil ETFs are in some ways in an odd position. When crude falls the market often rallies, improving investor sentiment. Bullish investors take the whole market up, including oil service ETFs like XLE and IYE. For the same reason, when the market is selling, energy ETFs sometimes sell, despite higher oil, which drives energy profits. This is short-term behavior. Ultimately, higher crude will likely push energy ETFs to trend higher.

But if the expectation is for higher oil, is this a good time to add to the oil ETF position? No. With oil down and oil ETFs still high, this looks like a very poor time to buy oil. A better time would be the reverse: higher oil and lower oil sector ETFs.

PORTFOLIO:

Looking forward, next week's Republican National Convention will dominate the news. Who knows how this will affect the market? It may provide a boost, as all the fat cats will be on television, smiling and waving the American flag. Of course, this means that there is a Risk of terrorism, which could hurt the market. Given this risk, I don't want to be too exposed during this time. I expect oil to bounce back, but the lower prices for crude and the uneasy political situation in New York require caution.

No change to the portfolio.

ETF	LW	TW	P/L
AGG	1	1	↑0.3%
XLE	2	2	↑0.5%
IBB	-1	-1	↓1.9%
Portfolio -0.2%			

LW: Last Week | TW: This Week | P/L: Profit/Loss

Week 10 : August 30 – September 3

QUESTION:
- *Which ETFs Are "Politically" Correct?*

INDICATORS:
- PD, EN, GD, SF

SITUATION:
The last week of summer, markets traditionally see some of the lightest trading of the year. This week was no different, and volume was low. News was dominated by the pageantry of the Republican National Convention in New York City. After last week's sell-off, oil futures stabilized, closing at $44.22 a barrel, up just over a dollar from last week. But this modest gain really pushed up energy ETFs. The chart below shows this week's returns of IYE, XLE, and IGE.

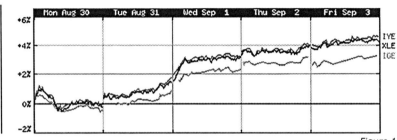

Figure 1

The relatively flatter lines in IGE are an indication that it is more

~45~

thinly traded than IYE and XLE. As the chart above shows, oil ETFs finished an already impressive summer with a 4% burst to new highs.

But as oil was climbing, look what happened to technology. Ouch.

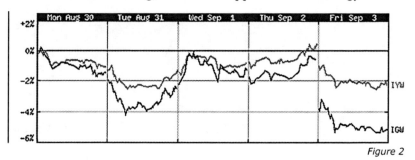

Figure 2

The chart above compares the semiconductors IGW with the more general technology fund IYW. Weakness in tech was due to disappointing earnings out of bellwether Intel (INTC). Though Intel represents just 8.2% of the holdings of the semiconductor ETF IGW, it dominates investor thinking on the semiconductor sector and influences the technology sector generally. IGW closed the week down 5.2%.

Utilities have been quietly outperforming all summer long, paying out dividends and moving upward on low volatility. (Utility ETFs have a 3-year beta of about 0.70).[1] The chart below compares the returns of the utilities sector ETFs XLU and IDU with the benchmark SPY.

---

1 Beta is a measure of risk. It involves specifically a fund's volatility in relation to the volatility of the general market. If a fund has a beta of 1, this means that when the broad market moves up, for example 2%, the fund will also move up 2%. If a fund has a beta higher than 1, the expectation is that it will move up more than the market. In the example above, if a fund has a beta of 2, and the market goes up 2%, the fund will be expected to go up twice as much as the market, or 4%. If the market were to drop 2%, a security with a beta of 2 would be expected to drop 4%. Utility funds have a beta of 0.7, which of course is less than 1 and means therefore that for every percentage point of the broad market's movement up or down, the utility fund is expected to move less than the market.

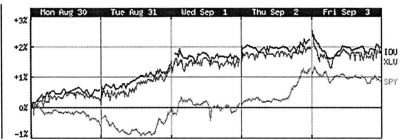

Figure 3

DISCUSSION:

As anybody watching the Republican Convention on television this week knows, this is Bush country. And, like it or not, this is the Bush economy. And the Bush economy is not about technology. It's about fear: terrorism, war, and oil. The favorite Bush sectors are defensive groups like oil, military contractors, and utilities. Unfortunately, there is currently no ETF for the defense contractors or security providers. But the utility sector is a Bush play. As the chart in Figure 3 above shows, the utility sector is up almost 2% on the week. It is also at new 52-week highs.

I like to buy when an ETF goes to new highs, especially in a leading sector, with a smooth chart. A smooth chart is usually less risky than a sharp move upward (after a stock shoots straight up), because of a single event like a drug approval. Utilities have risen slowly and steadily. The utility chart is very gentle, but up. This kind of chart points to further gentle upside – a nice new long to add to the portfolio.

Summer is over, and the portfolio managers are going to come back to Wall Street and look at the utility and oil sector charts and see how gentle and strong they look, and they will want to be long. Add the utilities (IDU).

PORTFOLIO:

Oil ETFs ended the week on an all-time high. It's time to be long and get even longer.

Technology goes lower and lower. I want to increase my short exposure to tech.

Nothing seems to be happening with the AGG trade. Maybe I should exit it simply to reduce exposure or boredom. Lousy idea. I have made mistakes in the past when looking for excitement rather than a good trade. There is plenty of excitement at a baseball game, a bar or even a casino.

⚠ *Never go to the market for excitement, or trade as if looking for excitement. This makes the market into a casino and the trader into a gambler. Gamblers always lose. At a minimum, all these trades cost commission, and this goes to the house. The best long-term result for anyone who trades for excitement is to lose the VIG, the commission, and most will lose a lot more – like everything.*

I want to keep AGG in the portfolio another week to remind myself that this is not about the love of the trade. It is not about excitement. It is not even about being right or wrong. It is about making money. Simple.

ETF	LW	TW	P/L
AGG	1	1	0%
XLE	2	3	↑4.0%
IBB	-1	-1	↑0.1%
IDU	0	1	
Portfolio +1.0%			

LW: Last Week | TW: This Week | P/L: Profit/Loss

Week 11: September 7 – September 10

QUESTION:
- *"Sell in May and Go Away," but Is Fall a Good Time to Buy?*

INDICATORS:
- SF, PD, EN, GD

SITUATION:
The markets are up, just about all of them. Traders returned to Wall Street from the long Labor Day weekend in a buying mood. The Republican National Convention ended without incident. The first week of fall, and suddenly the markets get serious again. Traders are thinking: Looks good, politics are OK, America is OK. Buying season: time to get long.

The bulls were also helped by the July trade balance, out this week. At -$50.1 billion, this is an enormous number, but it is an improvement over last month, and traders feared worse. Another positive factor was the August PPI (Producer Price Index), a key measure of inflation. The PPI came in Friday at -0.1%, meaning that inflation was slightly negative. This is remarkable given the rising price of oil, which is a key input for so many products. There is no question: Oil is getting more expensive. If inflation is stable, something else must be getting cheaper to offset the cost of higher oil.

Oil dropped slightly to $42.81 a barrel, the lowest weekly close in a month. But oil ETFs remained strong. The chart below shows that oil ETFs XLE, IYE, and IGE are outperforming the Dow Jones Industrial Average and the benchmark ETF that tracks Dow Jones Industrial Average proxy DIA.

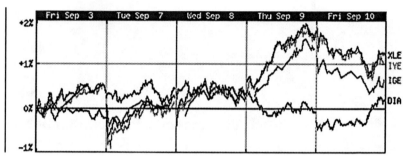

Figure 1

If earnings concerns held back the progress of technology ETFs over the summer, investors obviously decided to lay these concerns aside this week. Technology ETFs jumped ahead, outperforming the Dow. The chart below compares this week's returns of the Dow stocks, represented by DIA, with the overall market, represented by the Vanguard Total Stock Market (VTI), the technology-heavy QQQ, and the technology-specialized IYW.

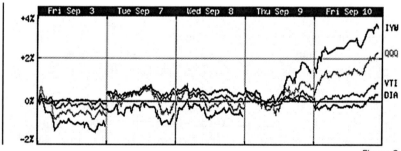

Figure 2

As the chart shows, although oil did well, with the broad markets improving and ending the week on a high, technology was the place to be.

DISCUSSION:

Lots of good news. Fall is here, and the bulls are running. Even oil ETFs stayed up this week, despite lower crude. Time to buy?

Not yet. Time to stay gloomy. The key chart is Figure 2 above. Despite the good news, the broad market was under water all week. The market perked up only on Thursday and then had a really good Friday on follow-through to Thursday's rally, including perhaps short covering. But two days do not a rally make. The trend continues to be lower tech, and there is not enough evidence here to reverse course and go long, especially in the volatile technology sector, which saw the most dramatic gains.

Energy retains leadership. Buying tech remains dangerous. Yes, this may be the beginning of a rally, but there is some buying expected in the beginning of fall. Before reversing course, confirmation of this apparent new trend is needed.

PORTFOLIO:

No change to the portfolio.

ETF	LW	TW	P/L
AGG	1	1	↑0.4%
XLE	3	3	↑0.9%
IBB	-1	-1	↓0.8%
IDU	1	1	↑0.2%
Portfolio +0.2%			

LW: Last Week | TW: This Week | P/L: Profit/Loss

Week 12: September 13 – September 17

QUESTION:
- *What Does the Transportation Sector Tell Us?*

INDICATORS:
- GD, TA

SITUATION:
The 10-day post-summer bull market took a breather this week, closing mostly unchanged. One possible reason: no big economic surprises. The chart above shows this week's flat performance of three of the DIA, QQQ, and SPY.

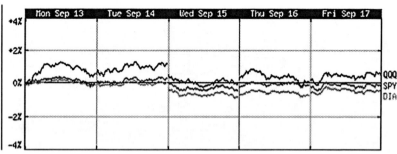

Figure 1

But if overall the market lacked momentum, some sectors continued to churn ahead. Investors liked the earnings optimism from key industrial company Ford Motor (F) this week. And the iShares Dow Jones Transportation Average (IYT), which seeks returns that correspond to the Dow Jones Transportation Index, hit new highs this week, despite a testy oil contract that moved higher again, hitting $45 a barrel amid concern that hurricane weather in the Gulf of Mexico could hamper supply. The chart compares the ETF proxy for the transportation average (IYT) performance this week to the benchmark MidCap SPDR Trust (MDY).

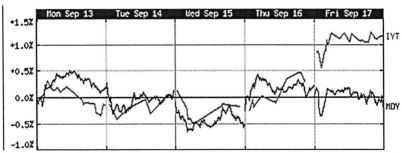

Figure 2

DISCUSSION:

Charles Dow developed the "Industrial Average," which at its inception held 12 blue-chip industrial stocks, and the "Rail Average," originally composed of 20 railroad companies, in order to make a broad assessment of market direction. These averages became what is known today as the Dow Jones Industrial Average and the Dow Jones Transportation Average. Today, the transportation proxy IYT holds few "rail" companies. It is comprised of companies like Fed Ex (FDX), United Postal Service (UPS), and Ryder.

According to the Dow Theory, the market is on an upward trend when either the industrial or transportation average advances to a new high, followed by the other. This week, IYT closed at a new 52-week high despite the high cost of oil, a key expense for transportation companies. This seems like a strong endorsement for the market.

However, as the chart below shows, though IYT has consistently made new highs, the more industrial broad market DIA, which tracks the Dow Jones Industrial Average, and the benchmark SPY have not followed along. They have been stagnant to lower.

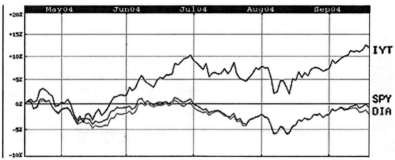

Figure 3

Without confirmation from the DIA, market direction according to the Dow Theory is inconclusive as to the direction of the market. Therefore despite the remarkable performance of the IYT given the sector's exposure to oil prices, nothing is conclusive. Perhaps the DIA is due for a rally. An upward move in the DIA would fulfill the conditions of the Dow Theory for a upward market.

PORTFOLIO:

No Change.

ETF	LW	TW	P/L
AGG	1	1	0%
XLE	3	3	↑2.8%
IBB	-1	-1	↓2.2%
IDU	1	1	↑0.5%
Portfolio +0.4%			

LW: Last Week | TW: This Week | P/L: Profit/Loss

Week 13: September 20 – September 24

QUESTION:
■ *Oil and Tech Redux: Does the Inverse Relation Maintain?*

INDICATORS:
■ EN, TA, GD

SITUATION:
Crude did well this week, moving to a record $48.88 a barrel Friday. High oil offset this week's better-than-expected durable goods number, and the broad market tumbled. The two charts below compare energy sector ETF XLE with the tech-heavy QQQ.

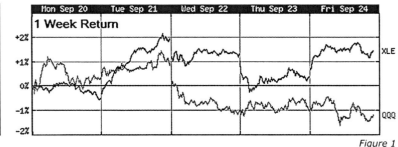

Figure 1

The chart above, compares the 1-week return of XLE and QQQ. The second chart, below, compares the returns of the XLE and QQQ over a 3-month period.

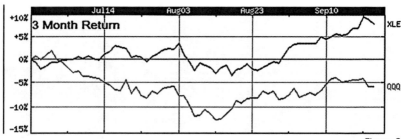

Figure 2

DISCUSSION:

As the two charts show, although the relationship is not perfect, technology and oil sectors continue to have a partial mirror relationship. Technology also looks good short side. The portfolio should probably focus specifically on what is working best: long oil and short tech. Always focus on what is easiest and working best. The easiest trade is the best trade. I'll increase the size of the oil long and tech shorts here.

PORTFOLIO:

ETF	LW	TW	P/L
AGG	1	1	↑0.5%
XLE	3	4	↑2.1%
IBB	-1	-2	↑1.9%
IDU	1	1	↓1.1%
Portfolio +1.0%			

LW: Last Week | TW: This Week | P/L: Profit/Loss

Week 14: September 27 – October 1

QUESTION:
■ *Is the Bull Back?*

INDICATORS:
■ SF, EN, GD, CC

SITUATION:
Despite rising oil, which settled at a new high of $50.12 a barrel Friday, the market was very kind to just about everyone in this first week of the third quarter. With national news dominated by the presidential debate, and few critical economic numbers out to provide direction, investors put aside worries, focusing instead on opportunities.

The new and still thinly traded materials sector made huge gains. The ETF that represents this sector is the Vanguard Material VIPERs (VAW). It seeks to match the performance of the MSCI US Investable Market Materials Index. VAW focuses on materials and holds companies like aluminum producer Alcoa (AA), gold and metals mining company Newmont (NEM), and diversified chemical companies DuPont (DD) and Dow Chemical (DOW). The chart below compares the weekly returns of VAW with two oil-sector ETFs, IGE and IYE. The flatness and the breaks in the VAW chart indicate that there the ETF is thinly traded. There are relatively fewer data points representing trades.

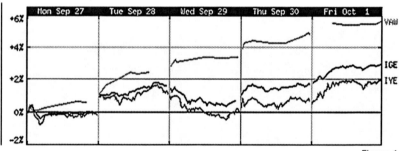

Figure 1

A strong report on August construction spending, which came in higher than expected Friday, pumped up REITs, undoing last week's losses in the sector. The chart below compares the weekly returns of the REITs IYR and RWR with the benchmark SPY.

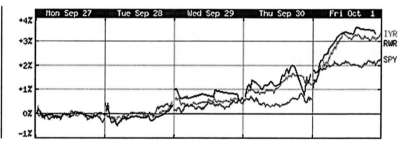

Figure 2

And even technology is strong. The chart below compares the week's return of the software index IGV, the semiconductor IGW, and a diversified tech fund, IGM, to the benchmark SPY.

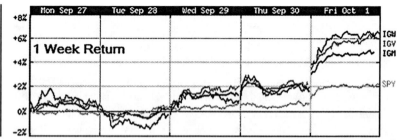

Figure 3

Internationally, there was good participation, Europe, Asia, and Latin America mostly outperforming U.S. benchmarks. The chart below compares this week's return of two Asian ETFs: iShares MSCI Japan Index (EWJ), and iShares MSCI Taiwan Index (EWT) with a European fund, the iShares MSCI EMU Index (EZU), Brazil's EWZ, and the U.S. benchmark SPY. The chart shows that international funds are outperforming the domestic benchmark SPY.

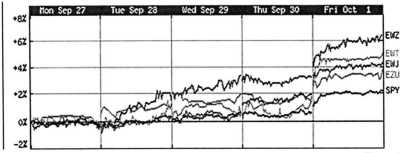

Figure 4

As investors channeled money into equities, bonds were the notable losers. They fell dramatically almost every day, all across the board, with longer-duration funds underperforming. The chart below compares the returns of three fixed-income funds, TLT, AGG, and LQD, with the benchmark SHY.

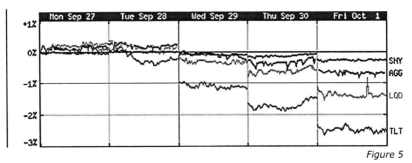

Figure 5

DISCUSSION:

It isn't just the old-line industry this week. The bulls have control of this market. On the long side, everybody is making money, from energy and materials to REITs to technology. The shorts are getting killed. Another good sign for equity valuation is the global participation. The under-performance of bonds is also bullish, as it indicates that investors are more risk-oriented and chasing profits rather than cozying up with bonds.

Technology remains the sticky point. What specifically was the anatomy of the technology rally? As the chart above in Technology boomed on Friday — especially Friday — hurting the shorts.

As the chart in Figure 3 above shows, this is an excellent return for tech. But all tech shorts shouldn't necessarily panic — yet. Pulling up the 6-month chart may help to determine if this rally at last is real:

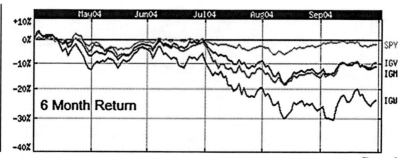

Figure 6

The 6-month chart shown in Figure 6 above shows what the 1-week chart does not. The 6-month chart is the mirror image of the 1-week return in Figure 5. Review of figures 5 and 6 together show that technology's performance on a 1-week basis is phenomenal, but the 3-month basis dismal. Buying technology still remains a contrarian bet.

Nevertheless, given this momentum, it may be a good time to try to grab a bargain here. IGW is down 25% in three months. For a contrarian bet in technology, this is the one to buy. A significant reason for caution: Oil prices are hitting new highs, and higher crude has often meant that the market stagnates and tech goes lower. Although it is great to see the whole market rally, what is the leadership sector? In the late 1990s, technology moved up, while old-line industries stagnated. Is this tech-driven market back? It remains way too early to bet this way. This would be calling a change in trend, and this is always difficult and dangerous. With oil continuing to be strong, and defensive sectors like materials moving to highs, the re-emergence of technology as a leadership sector — at least in the short run — seems doubtful.

PORTFOLIO:

One reason for the strong week may be the seasonal factor.

▲ *At the end of each quarter, there is a lot of portfolio rebalancing known as "window-dressing," as fund managers sell losers and buy winners, so that when they report holdings to their clients, clients will be pleased that funds are invested in securities that are performing well. There may be opportunities to take advantage of this window-dressing phenomenon by buying strong stocks or selling weak stocks ahead of fund managers' last-minute purchases.*

In this case, some of the selling in IGW may be because portfolio managers are getting rid of semiconductors due to under-performance. Likewise, in the beginning of each quarter, managers adjust portfolios for the quarter to come.

Among the technology sector, semiconductors look the most oversold. If the market improves next week, everyone will have sold, and IGW will go up. Because of the short IBB position, I am as of now weighted to the short side – to benefit from bad news. So a bet on IGW would be a partial hedge. The chart looks as if it is also starting to climb. I'm ready to add IGW.

With bonds falling and the market climbing, it is time here to get out of AGG.

ETF	LW	TW	P/L
AGG	1	Closed	↓0.5%
XLE	4	4	↑2.3%
IBB	-2	-2	↑2.0%
IDU	1	1	↑1.7%
IGW	0	1	
Portfolio +1.4%			

LW: Last Week | TW: This Week | P/L: Profit/Loss

Part II, Q4 2004

~

Week 15 - Week 27

Week 15: October 4 – October 8

QUESTION:
- *How Important Is the Monthly Employment Report?*

INDICATORS:
- GD, EN, MR

SITUATION:
Nonfarm payrolls came in at 96,000 versus the 140,000 expected. Oil spiked, closing the week at another new high of $53.31 a barrel. The chart compares the large-cap fund VTI to a mid-cap fund MDY, and a small-cap issue that tracks the Russell 2000, IWM.

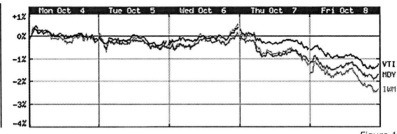

Figure 1

As the chart shows, as long as ETFs remained flat, all three traded together. Small-, medium- and large-cap ETFs traded together but moved apart Thursday and Friday.

Friday's negative employment report had a dramatic impact on all fixed-income ETFs. The chart compares the returns the long bond fund TLT with the mid-term government ETF IEF, and the short-term fund SHY.

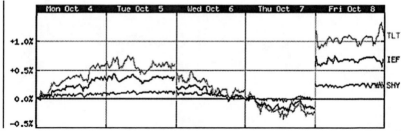

Figure 2

Energy ETFs followed higher crude prices to new highs again this week. As a group, energy ETFs have had consistently high returns without a single down week in two months.

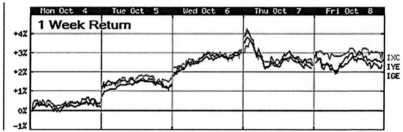

Figure 3

The chart above shows the 1-week return of IGE, IYE, and IXC, The chart below shows the 3-month return.

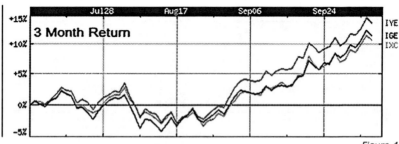

Figure 4

The sensitive technology sector fell lower as oil climbed. IGW, the semiconductor ETF, closed the week down almost 5%. The chart below compares IGW with Software (IGV), and Networking (IGN).

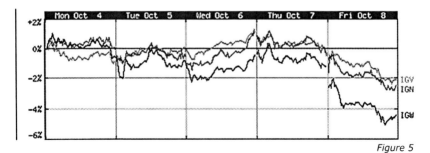

Figure 5

DISCUSSION:

From the perspective of market movement, the most important economic data released by the government is the Employment Report. September employment missed by a mile, nearly a third lower than expected. This kind of disappointment really shakes the market.

The most clear sign of the market's disappointment is the sharp move upward in fixed-income shown in Figure 2 above. Fixed-income investors watch macroeconomic indicators particularly closely because signs of a weak or weakening economy impact bond prices. Typically, a weak employment discourages the Federal Reserve from raising rates and encourages bond investors. Good macroeconomic news often has the reverse impact, causing bonds to sell sharply. Changes in market sentiment are typically most important to the sensitive longer-term bond portfolios such as TLT.

As the chart in Figure 1 shows, small-caps are often disproportionately negatively affected by negative macroeconomic news such as the employment report. High beta technology funds like those in Figure 5 can also be expected to sell off sharply.

Finally, a poor employment number hinting at a poor economy makes it harder to think of a return to playing offense and the technology sector leadership. Figure 5 shows the high beta technology ETFs taking a big hit when the market outlook turns sour. Until the bull market is back, any long tech position is potentially very dangerous. The long-term trend that we are seeing week after week is higher oil and lower tech. Until the trend changes, buying technology is for suckers.

PORTFOLIO:

I am getting out of the semiconductor position. After so sudden and sharp a loss in IGW, what have I learned? I've learned that I have to be careful about the employment number on technology. I've learned that semiconductors move very fast. A more broad-based ETF with tech exposure, such as IGM or QQQ, might make a better choice than IGW for a broad tech bet. I'll take a loss in the IGW position, hopefully wiser, or at least more patient, next time.

By contrast, the oil sector continues to be profitable. Oil ETFs are at new highs for the eighth week in a row, and by all indications, going higher. I'm going to add more oil to the portfolio.

I am also going to add to the profitable IBB short position on this drop.

ETF	LW	TW	P/L
XLE	4	5	↑2.1%
IBB	-2	-3	↑4.0%
IDU	1	1	↑0.9%
IGW	1	Closed	↓4.2%
Portfolio +1.7%			

LW: Last Week | TW: This Week | P/L: Profit/Loss

Week 16: October 11 – October 15

QUESTION:
- *Randomness: Is There Always a Reason?*

INDICATORS:
- EN, TA, IP, CC, MR

SITUATION:
Oil closed this week at a new high of $54.93 a barrel and continues to show few signs of weakening. Yet while oil prices were making new highs, investors took profits in energy-sector ETFs sending them sharply lower. The energy sector significantly underperformed the broad market for the first time in weeks. The chart below compares the returns of XLE and IGE to the SPY.

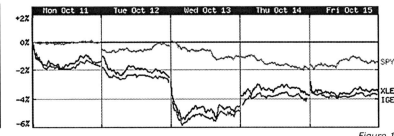

Figure 1

Everybody seems worried. There is no refuge from high oil. Record-high oil prices offset positive economic data this week, such as an impressive 1.5% increase in September retail sales. High oil is also having an impact on the U.S. trade deficit, which widened 6.9% to $54 billion in August, the second-highest ever. Internationally there is also concern. Most markets closed with significant losses. The chart below compares the returns of the SPY, the Europe S&P 350 Index (IEV), Japan (EWJ), and the Emerging Markets Index (EEM).

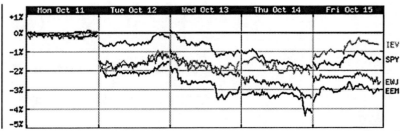

Figure 2

But as has often been the case this year, bad news for equity was good news for bonds. Bond investors were energized by the initial jobless claims, which increased to 352,000, sending the number of U.S. unemployed to the highest level since February. Many bond investors are reasoning that a sluggish economy as reflected by high unemployment and expensive oil will stay the hand of the Fed and that interest rates will not rise quickly. The result is a powerful rise in fixed-income values. The chart below shows the remarkable mirror return of the bond benchmark TLT and the equity benchmark SPY.

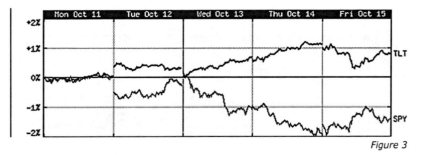

Figure 3

DISCUSSION:

"We underestimate the share of randomness in everything" writes

Nassim Taleb.[1] Following Taleb, any useful discussion on randomness and the markets is necessarily a long one. Even defining randomness is not simple. And randomness is also different from unpredictability. But practically speaking, I think for the purposes of a trader both the unpredictability and randomness of the market are overstated. It can however be useful for a trader to remember that not all market movement signals, or signifies.

The mirrorlike chart in Figure 3 above showing the relationship between stocks (SPY) and bonds (TLT), for example, is clearly not random and clearly signals. Figure 3 signals that SPY and TLT are clearly trading in opposite directions. The shape of this chart indicates that investors are buying bonds whenever they lose confidence in equity. Whenever the broad market falls, bonds become attractive. Investors sell bonds when confidence returns.

One of the themes of this book is the relation between crude prices and energy ETFs. Over the past six months, energy ETFs have consistently followed higher oil prices to gains. Although the price of crude and the price of energy ETFs are not perfectly correlated, they are highly correlated. This week, energy prices made new highs, but oil service ETFs sold off with the rest of the market. The word on the street is that unlike previous highs, the high cost of crude seems to have attained a threshold level, where rather suddenly everyone is concerned about energy and how record prices will affect growth in economies worldwide.

With this as the headline story, global markets are selling, but why are oil service ETFs also selling? The logic here is that if oil goes higher, this will eventually slow the economy and choke off demand for oil. It can be helpful at times to call a divergence for which one can find no explanation "random." I have no explanation for why oil service ETFs are lower on these record high crude prices. When constantly looking for explanations, it is worth remembering that there may be no explanation that fits and helps to formulate a plan and strategy. Despite the new publicity, nothing fundamental has changed this week. I would like to increase the XLE exposure, with the expectation that energy ETFs selling this week on record high oil prices will reverse next week.

---

1   Taleb, Nassim, *Fooled by Randomness*. New York: Random House, 2004, p. xli.

PORTFOLIO:

I would like to add to the oil position here on what looks like "random" noise. But I need to keep discipline and therefore will instead reduce my position as I do when I lose more than about 1%. But I will reduce the XLE by as little as possible.

Biotech continues to sell. The biotech short can be increased. I'm shorting QQQ again.

ETF	LW	TW	P/L
XLE	5	4	↓2.8
IBB	-3	-4	↑1.6%
IDU	1	1	0%
QQQ	0	-1	
Portfolio -0.3%			

LW: Last Week | TW: This Week | P/L: Profit/Loss

Week 17: October 18 – October 22

QUESTION:
- *What Does the Fed Model Tell Us?*

INDICATORS:
- EN, TA, FA

SITUATION:
Despite the Federal Reserve's 25 basis-point rate hike Tuesday, yields on the 10-year Treasury bond are below 4% for the first time since March. Low yields means high prices for bonds. The 6-month chart below shows the strength of the intermediate-term bond fund IEF, trading inversely to the benchmark SPY.

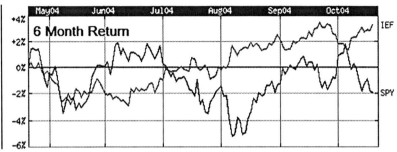

Figure 1

Can these low yields be sustained in a rising rate environment?

DISCUSSION:

The Fed model, so named because it was developed by the Federal Reserve, attempts to deal with the relationship between bond prices and stock prices by comparing the yield on the benchmark 10-year Treasury note with the yield earned from dividends paid by holding the S&P 500.

The model works with the assumption that investors are willing to assume the risk of owning stock as compared with bonds if the rewards are high enough. What kind of return can investors get from an investment in the S&P 500 as compared with Treasury bonds for assuming the additional risk of owning equity rather than bonds?

A *The Fed model gives an indication of how attractive a Treasury bond yield is relative to the yield an investor can receive from dividends paid by owning the S&P 500. The assumption is that investors are interested in buying stocks only when their expected earnings exceed returns from risk-free fixed-income securities.*

What follows here is a quantitative discussion of the current Fed model. The Fed model centers on the following ratio:

S&P 500 annual forward earnings/annual yield on 10-year Treasury notes

Thomas Weisel Partners estimates the 12-month forward earnings on the S&P 500 index at $69.94 through December 2005. This number is comparable to J.P. Morgan Chase's $69.43 estimate. The current yield on the 10-year Treasury note is approximately 4%.

To calculate the minimum fair value of the S&P according to the model, we can divide Thomas Weisel's estimate of $69.94 by the current Treasury bond yield of 4%. The result is $1,748.50. The idea here is that investors should not pay more than this because they can get an equivalent return from risk-free Treasuries. With the S&P currently trading at $1,106, investors are paying approximately 37% less for stocks than for bonds. This discount (also known as an "equity risk premium") is very favorable in historical terms.

Looking at the numbers, there have been few instances in recent history when the S&P 500's expected earnings have so greatly exceeded fixed-income yields. In other words, the Fed model shows that stocks appear to be extremely cheap compared to bonds. The table below shows key points when the S&P was at a positive or negative discount to the Fed model's minimum fair value:

	Forward Earnings	10-Year Treasury in %	Fed Model Minimum Fair Value	S&P 500	% discount of S&P to Fed Model
Jan 1990	28.92	8.21	352.3	340.0	-3.5
Jan 1995	37.67	7.78	484.2	465.3	-3.9
Jan 2000	58.09	6.66	872.2	1425.6	63.4
Jan 2001	59.51	5.16	1153.3	1335.6	15.8
Jan 2002	52.27	5.04	1037.1	1140.2	9.9
Jan 2003	54.84	4.05	1354.1	895.8	-33.8
Jan 2004	61.98	4.15	1493.5	1132.5	-24.2
Oct 2004	69.94	4.00	1748.5	1106	-36.7

Source: Thomson Financial, Prudential Securities, Bloomberg
Table 4

In the table above, a negative percentage suggests that the S&P may be underpriced, while a positive percentage suggests that the S&P may be overpriced in comparison with the model.

The numbers in the table above were computed by the formula:

$$SP/(E/Y)-100$$

where SP is the S&P 500 index and E is forward earnings, Y is the 10-year Treasury yield, minus 100. Thus, the current comparison ratio is

$$[1,106/(69.94/.04) = 63.3\%]-100\%, \text{ or } -36.7\%.$$

This number is shown in the lower-right corner square in Table 3 above.

In 1990 and 1995, fair value of the S&P, as given by the Fed model, was close to the level where the S&P was trading. These both would have been fine times to enter the market for a multiyear period. In 2000, the Fed model predicted that the S&P was wildly overvalued, which in retrospect it was. In January 2001 and 2002, the Fed model's minimum fair valuation was only slightly higher than the S&P, and indeed markets continued to be soft. By January 2003, the S&P was trading significantly under the Fed model's valuation, and, in retrospect, the S&P was undervalued. The S&P's

discount to the Fed model valuation continued throughout 2004, and in October, the discount reached a level with few historical precedents.

Of course when Treasury yields are historically or exceptionally low, as they seem to be now, the Fed model formula generates a high figure. Many observers predict that interest rates will be pushed up by U.S. government borrowing in the face of huge budget deficits. If this happens, this will have the effect of pushing down the Fed model's minimum fair value of stocks. For instance, if Treasury yields were to climb to 5% immediately, the S&P's market price would trail the Fed model's minimum fair valuation by 21%, not 37%. Stocks would not appear quite so cheap. But they would still be a good value.

The Fed model may be a better long-term valuation tool than a short-term market-timing tool, as differences between fair value as given by the model and the S&P Index can endure for longer periods, such as in 2003 and 2004. Nonetheless, it stands now at an historical extreme, and should at a minimum be a reminder about the dangers currently of being short. There looks like a lot more opportunity to the upside.

While there may be good reasons for being short individual issues – such as, say, Martha Stewart (MSO), because an investor believes that her brand will not survive if she is given a jail sentence, or Krispy Kreme Doughnuts (KKD), because the Atkins carbohydrate-free diet will hurt sales, ETF investors are involved in larger sector- or country-wide trends. The reason to short an ETF is that the market (or a big chunk of the market) is going lower. According to the Fed model right now, stocks are cheap, so it does not look like a good time to put on short positions.

PORTFOLIO:

The thing to do here is to position for a rally. But given the potential for a terrorist attack and the direction of this market, I am still nervous about shorting TLT. I am also nervous about buying technology only to catch a falling knife.

I find these levels as described by the Fed model persuasive and believe after looking at these numbers that being short here is more dangerous than being long. As a result, I want to cover up the short biotech and cubes position by buying back IBB and QQQ. And also, buy the broad market SPY, which tracks the S&P 500, upon which the Fed model is based.

ETF	LW	TW	P/L
XLE	5	5	↑0.9%
IBB	-4	Closed	↑2.0%
IDU	1	1	0%
QQQ	-1	Closed	↓0.8%
SPY	0	1	
Portfolio +1.2%			

LW: Last Week | TW: This Week | P/L: Profit/Loss

Week 18: October 25 – October 29

QUESTION:
- *Are Utilities a Buy?*

INDICATORS:
- EN, GD, IP, FA

SITUATION:
Oil closed at 51.78 a barrel, lower on a weekly basis for the first time in a month and a half, and down more than $3 from last week's record close of 55.17. Explanations for lower oil ranged from a cooling of the Chinese economy, after its central bank raised interest rates, to hedge funds moving to unwind substantial long positions.[1]

In any case, the broad market celebrated, shrugging off election uncertainty, concern about possible terrorist attacks, and a disappointing third-quarter GDP number, which came in over half a point lower than expected, at 3.7%. Lower oil is the bigger news. After making new lows for the year just last week, the Dow Jones Industrial Average proxy (DIA) once again rebounded to more than the psychologically important 100 this week, amid a broad rally. The chart below compares the DIA with the QQQ.

---

[1] When no other explanation is plausible, hedge funds are often cited as a reason for market movement.

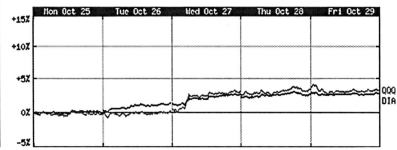

Figure 1

Good news in the broad market has not, however, meant that investors have abandoned defensive plays. The utility sector also continues to do very well, and is still a little under the radar. Already up 20% over the last 12 months, utilities moved to new highs this week, as investors continued to be attracted to their yields. The two charts below show the returns of three utility ETFs: XLU, IDU, and VPU. The upper chart shows this week's return. The lower chart shows the 1-year return:

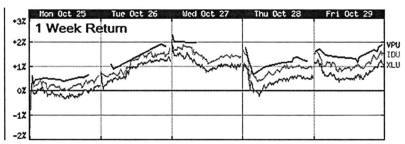

Figure 2

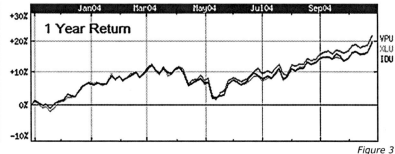

Figure 3

The smooth curve on VPU as compared with XLE and IDU

indicates that VPU is relatively thinly traded and therefore illiquid.

Investors also bought REITs, another defensive yield-paying sector up more than 20% in the last 12 months and now at new all-time highs. The two charts below compare the 1-week and 1-year returns of three REITs: ICF, IYR, and RWR.

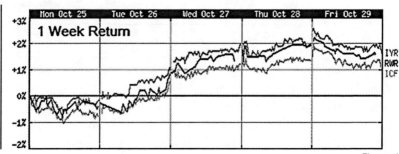

Figure 4

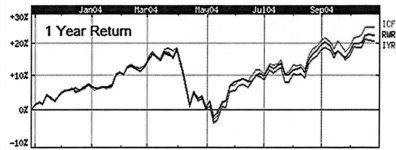

Figure 5

As the two charts above show, these REIT ETFs have very similar performance on both a weekly and yearly basis.

DISCUSSION:

The long-term utility chart in Figure 3 above is an investor's dream. Traditionally thought of as low-risk securities, utilities are historically popular for their ability to deliver predictable cash dividends. But in the current lackluster market, utilities have become the new growth stocks. Over the past year, while the broad market has stagnated, utility ETFs have been on a tear, up 20%. This is remarkable because utilities have among the lowest beta of any ETF, typically under 0.7, which means that they are expected to have less market risk than other ETFs.

Utility ETFs over the past year have been the beneficiaries of several important trends. First, and probably most importantly, utilities have benefited from a decline in long-term bond yields. On a comparative basis, the reliable dividends utilities pay look more attractive in a low-yield environment where treasuries are paying near 4% than when Treasuries are paying 6% or 8%. Second, utilities have benefited recently from an increase in mergers and acquisitions activity in the industry. Third, utilities are benefiting from a regulatory-friendly Bush administration. Fourth, utilities are boosted by an environment of broad market stagnation and investors looking for defensive plays and relative safety. Utility ETFs have also benefited for a fifth, technical reason − simple momentum in the sector.

The REIT sector also benefits declining long-term bond yields. They also have a low beta and often do well in an environment of broad market stagnation. And like utilities, REITs currently have momentum. On a weekly basis, the utility chart in Figure 2 and the REIT chart in Figure 4 look similar. But comparing the 1-year utility chart in Figure 3 above to the 1-year REITs chart in Figure 5 above, whereas REITs suffered a significant sell-off in April, utilities moved steadily upward. Because of the volatility of the REIT sector, the utility sector trade looks easier.

The main risk in buying utilities now is that if Bush is not reelected and that Kerry, if elected, will want more significant regulation. If this happens, this could eliminate a reason for owning utilities.

PORTFOLIO:

One option would be to wait until after the election before buying utilities. Instead, I'm going to add more utilities ETFs to the portfolio on this high and then adjust according to the result of the election and if policy directives change.

With oil continuing to fall, I want to lighten up slightly on oil. On a 1-week basis, TLT looks weak.

Following the Fed Model described in the previous chapter, I want to initiate a short position here in TLT.[2] I'd like to buy more

---

2   Unfortunately, there is no TLT available to short. So I am selling five

SPY, but this has been a strong move already. I don't want to be too aggressive. I'll wait a week before adding to the position.

ETF	LW	TW	P/L
XLE	5	4	↓0.9%
IDU	1	2	↑2.6%
SPY	1	1	↑3.1%
TLT	0	-1	
Portfolio +0.9%			

LW: Last Week | TW: This Week | P/L: Profit/Loss

September at-the-money calls on TLT. For the sake of simplicity, this options trade is represented as a -1 short position in the portfolio.

Week 19: November 1 – November 5

QUESTION:
- *What Does a Bush Victory Mean for the Market?*

INDICATORS:
- PD, EN, GD, CC

SITUATION:
Just about everything went right for ETF investors this week. The market was awash in news positive to the broad market: a decisive Bush victory, declining oil prices, and an improved employment outlook, with almost twice as many jobs created as expected. The chart below shows how the DIA and QQQ traded on the election news.

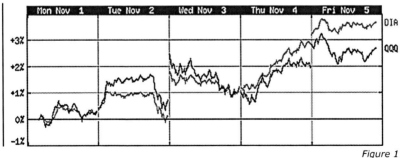

Figure 1

The chart above shows that both the DIA and QQQ sold off late

~81~

Tuesday, as exit polls suggested that Kerry was in the lead for the presidency. The market closed before the final result was announced and reopened higher Wednesday morning after Bush was confirmed the winner. On Thursday and Friday, the market rallied, as oil fell and the employment picture improved.

One area of concern is the continuing fall of the dollar. The dollar is at an all-time low against the euro and falling against the yen and peso. Dollar weakness is positive for foreign ETFs. The chart below compares Japan (EWJ), Europe (EZU), and Mexico (EWW).

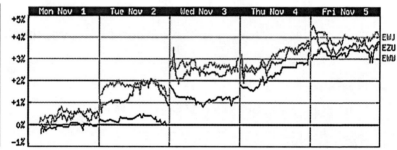

Figure 2

Comparing the returns in Figure 1 and Figure 2 above shows that, despite the celebration from the Bush victory domestically, these international ETFs outperformed domestic markets by 1% to 2%.

The chart below shows the inverse relation of the equity benchmark SPY and the fixed-income benchmark TLT.

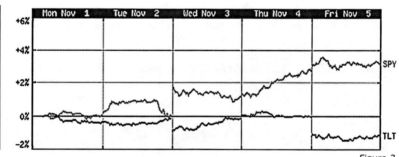

Figure 3

Wednesday's gain in the SPY (and loss in the long bond fund TLT) reflects the election results. Friday's further gain in the SPY (and further loss in the bond portfolio TLT) is due to the positive employment number. The sentiment is that robust employment would do nothing to discourage the Fed from raising rates another 25 basis points next week.

Some of the factors putting fixed-income under pressure may also impact REITs. REITs have had a phenomenal year but carry interest-rate risk and are historically popular with investors seeking alternatives to conventional equity. But REIT ETFs have not yet seen any significant sell-off, and indeed are trading at their highs. This week, REITs followed the general market higher. But on Friday, REITs, like fixed-income, took it on the chin. The chart below shows this week's return of two REIT funds, RWR and ICF.

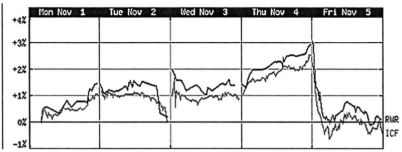

Figure 4

DISCUSSION:

Now that the election is over without a terrorist attack, the market, which has been worrying about this for the last six months, may finally move up. The SPY-TLT chart in Figure 3 above represents the fulfillment of what the Fed model predicted. SPY looks set to cross to a new high. I want to be long when that happens and want to increase the size of the Fed model trade: Buy more SPY, and short more TLT.

Twice as many jobs created as expected? This outstanding employment result is almost too good to be true, coming as it does right after the election. But why argue? This is not the time to be sector-specific on the long side. The broad market has done nothing all year, the celebration will be universal. So I'll buy the cubes. Oil finally is taking a back seat to overall optimism, and in fact falling.

As has often been the case this year, stocks and bonds moved in opposite directions. Bond prices fell as the week passed without a terrorist incident and as uncertainty surrounding the results of the election disappeared.

PORTFOLIO:

With the Bush crowd elected, the chance of Democrat Kerry-style utility regulation is nil. Utilities ripped to all-time highs. Time to get longer utilities.

Oil is below $50. Although this is a Bush sector, I am a little worried about my energy profits and want to reduce my exposure to oil. Good employment numbers are already out. The easiest way to juice the market now is lower oil. I want to cut the position size slightly.

ETF	LW	TW	P/L
XLE	4	3	↑0.9%
IDU	2	3	↑2.5%
SPY	1	2	↑3.6%
TLT	-1	-2	↑1.4%
QQQ	0	1	
Portfolio +2.0%			

LW: Last Week | TW: This Week | P/L: Profit/Loss

Week 20: November 8 – November 12

QUESTION:
- *What Is the Technical Outlook for Fixed-Income ETFs?*

INDICATORS:
- TA, PD, BY

SITUATION:
The post-election Bush rally continues. Broad market ETFs are at 52-week highs. Though most sectors did well, overall small-cap and mid-cap funds very significantly outperformed large-caps.

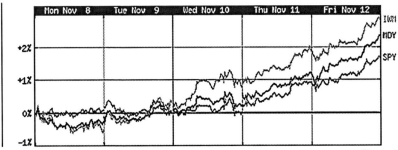

Figure 1

The chart above compares a small-cap fund, IWM, with a mid-cap fund, MDY, and the large-cap benchmark SPY.

Also helping the market was lower oil. Oil fell all week, settling at $47.32 a barrel. The dollar continued to drift lower.

The prospect of four more years of Bush has set a fire under utilities. They outperformed the broad market and are trading on their highs. The chart below compares the return of IDU and XLU.

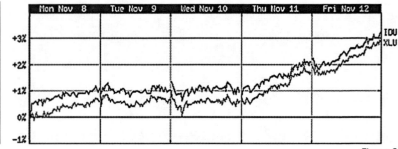

Figure 2

REITs were among the best performers on a 1-week basis, despite a rising rate environment. ICF was up over 5% for the week, 2.5% on Friday alone. The chart below compares the returns of ICF with RWR, and the Vanguard REIT Index VIPERs (VNQ).

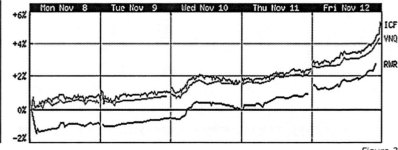

Figure 3

What does this environment suggest about fixed income?

DISCUSSION:

In the first 10 months of 2004, fixed-income ETFs kept pace with the S&P 500 and outperformed the Nasdaq. But will this trend continue? Evidence of weakness in fixed-income is mounting:

1. Rising broad market

As the chart in Figure 1 shows, the broad market is up. This week's outperformance of small- and mid-cap ETFs is a further bullish sign. It suggests that investors are willing to assume the typically higher risk of owning small companies.

With uncertainty of the presidential election and the Federal Reserve meeting behind them, investors may no longer be trading on fear. The economic optimism often reflected in market strength can be a bad sign for bonds. When the overall economic picture improves, investors become more aggressive. Many turn away from bonds in favor of equity.

2. Rising interest rates
3. Lopsided Fed model (See Chapter 7)
4. Technical deterioration

The chart below shows a 2-year price chart on TLT:

Figure 4

The chart suggests periods of sharp decline followed by a slow ramp upward. But in the 2-year period shown in the chart, each successive move upward is weaker than the previous one, and each successive move downward is stronger than the previous one.

It may be that TLT's regression from near 90 to about 87, where it traded recently, is unimportant and simply a slight pull-back on its way to the 90 mark and beyond. Even in this month, TLT may move higher than its March high. If this happens, the trend of lower highs and lower lows would be broken. For now, the trend persists.

Just as important, the chart on TLT is not unique. All longer-term bond funds have similar charts: lower highs and lower lows. IEF, which holds mid-term 7- to 10-year bonds and LQD, which holds corporate bonds, are no less suggestive than TLT of this pattern.

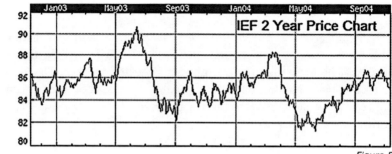

Figure 5

Figure 6

PORTFOLIO:

Because of the long maturity of the bonds held in TLT and its concomitant volatility, TLT will likely be more affected than IEF or LQD by rising yields. Add to the TLT short position here. Increase the SPY and QQQ positions on this strength.

ETF	LW	TW	P/L
XLE	3	3	↑0.5%
IDU	3	4	↑3.1%
SPY	2	3	↑1.3%
QQQ	1	2	↑1.8%
TLT	-2	-3	0
Portfolio +1.3%			

LW: Last Week | TW: This Week | P/L: Profit/Loss

Week 21: November 15 – November 19

QUESTION:
- *With Election Euphoria Over, Where Will the Market go Next?*

INDICATORS:
- EN, GD, FD, CC

SITUATION:
The election euphoria failed, and the price of oil resumed its upward climb, closing Friday at $48.89 a barrel. The Producer Price Index (PPI), an inflation benchmark, jumped 1.7%, making another Fed rate hike in December seem certain. The dollar lost ground helping international funds. The chart below compares the return of DIA with the Japan index (EWJ), and the European fund EZU.

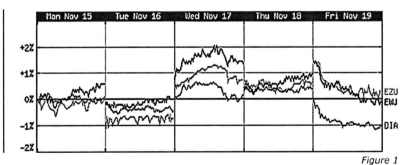

Figure 1

As the chart shows, foreign benchmarks outperformed the

~89~

domestic DIA. Friday's sell-off reflects the jump in PPI and concern about inflation.

An interesting story was the price movement in fixed-income. On Wednesday, investors focused on the rising equity market and a weaker dollar, arguing that demand for dollar-denominated equity was strong and foreign central banks would be forced to buy U.S. Treasury securities in order to hold the dollar up and thereby maintain favorable exchange rates for their exports. This sent bonds higher. But Greenspan's remarks Friday suggesting that foreign banks might tire of propping up the dollar changed that thinking, and investors reversed – selling bonds. The chart below shows this week's roller-coaster ride in the long bond fund TLT, the corporate bond fund LQD, and the mixed fund AGG.

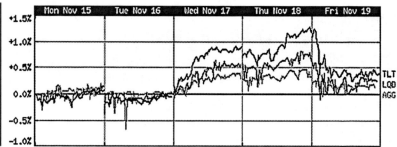

Figure 2

Concern about rising rates took a bite out of REITs. REITs continue to trade near all-time highs but have become increasingly volatile as investors try to gauge companies' exposure to rate changes. The chart below shows the returns of RWR, ICF, and IYR.

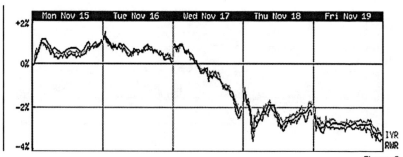

Figure 3

DISCUSSION:

The short REIT trade would have been right, but increasing the short bond position was a bad idea. Before Greenspan stepped in Friday, TLT was up over 1%. That is a big move. There is really tremendous appetite for bonds. On the other hand, the Fed model remains persuasive, as does the technical situation described last week. The risk here is that if this economy continues to stumble, people are going to step up and buy bonds.

In contrast to fixed-income, oil is giving me no trouble.

▲ *This is what any trader wants – not the charts that zigzag, but the slow and steady charts – higher highs, higher lows, followed by higher highs followed by higher lows – the easy trade.*

Although oil was down Monday and Tuesday, oil ETFs show the higher-highs, higher-lows trend for the rest of the week, ending the week on a high. The 1-week return and 6-month return of three energy sector ETFs: IYE, IGE, and XLE, follows.

Figure 4

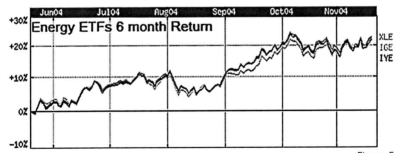

Figure 5

From the charts in Figure 4 and Figure 5 above, it looks as if oil has been consolidating and that it is going to break out and go higher. There seems to be no reason at all not to stay long energy and oil service ETFs. Stay long and get longer. Buy XLE.

PORTFOLIO:

ETF	LW	TW	P/L
XLE	3	4	↑1.3%
IDU	4	3	↓2.2%
SPY	3	2	↓0.8%
QQQ	2	2	0
TLT	-3	-3	↓0.4%
Portfolio -0.8%			

LW: Last Week | TW: This Week | P/L: Profit/Loss

Week 22: November 22 – November 26

QUESTION:
- *What Happens on Thanksgiving Week?*

INDICATORS:
- SF, CC, EN

SITUATION:
The broad market edged slightly higher this short Thanksgiving week. Oil rose, trading to just under $50 a barrel Friday, as traders worried about supply going into winter. October durable goods came in almost a point lower than expected. But the big news was the decline of the dollar. The dollar broke down this week, trading to a 5-year low against the yen and an all-time low against the euro. The euro bought $1.33, a 3% increase over last week.

Foreign ETFs benefited from the dollar decline. But investors began to worry that a weaker dollar would erode purchasing power domestically and therefore hurt foreign firms exporting to U.S. markets. Nonetheless, some of the best returns this week came from foreign ETFs. Brazil's EWZ had a great week, up 4.3%. The Europe 350 Index (IEV) was up 2.8%, a new high. By contrast, Japan's EWJ ended the week virtually unchanged, up just 0.1%.

Energy ETFs followed oil to all-time highs. The table below shows the 1-week and 52-week return of XLE, IXC, IGE, and IYE.

TICKER	1-Week Return	52-Week Return
XLE	3.4%	47.2%
IXC	3.4%	41.1%
IGE	3.3%	36.6%
IYE	2.8%	44.8%

Table 5

The utility sector moved to new highs this week. Utilities continue to be popular with investors because they provide excellent yields and are thought to be defensive. The table below shows the 1-week and 52-week returns of the three utility ETFs IDU, XLU, and VPU.

TICKER	1-Week Return	52-Week Return
IDU	2.8%	22.37%
XLU	2.6%	23.31%
VPU	2.8%	N/A

Table 6

DISCUSSION:

⚠ *The market usually goes up every Friday half-day after Thanksgiving.*

Energy and utilities continue to look like the hottest area. Add to both on these new highs. Otherwise, no change.

PORTFOLIO:

ETF	LW	TW	P/L
XLE	4	5	↑3.5%
IDU	3	4	↑2.1%
SPY	2	2	↑0.8%
QQQ	2	2	↑1.3%
TLT	-3	-3	0%
Portfolio +1.2%			

LW: Last Week | TW: This Week | P/L: Profit/Loss

Week 23: November 29 – December 3

QUESTION:
■ *What Does Counting Tell Us?*

INDICATORS:
■ EN, CC, GD, MR

SITUATION:
Oil collapsed, closing Friday at $42.50 a barrel, down over $7 from last week's close. Oil moved lower on increased inventory, and warmer weather and a resumption of supply in the Gulf of Mexico, which had been off-line due to hurricanes. Energy ETFs were some of the worst performers, selling Wednesday and Thursday. Energy recovered partly on Friday. The chart below shows XLE and IXC.

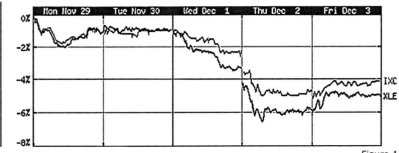

Figure 1

The broad market rallied on lower energy, but not as much as it might have given the severity of the fall in crude. The broad market was held back on worries about the continuing weakness of the dollar, which once again went to new lows this week against the euro and yen. Domestic small-caps outperformed mid-cap and large-cap issues. The chart below compares the large-cap DIA with the mid-cap MDY, and the small-cap iShares IWM.

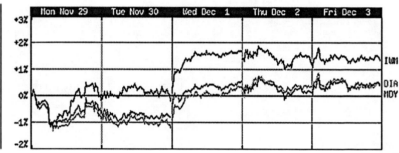

Figure 2

Economic news was weak. November nonfarm payrolls came in Friday at barely half the expected 220,000, an anemic 112,000. The news gave new life to bonds, which had sold off dramatically during the week as the broad market moved higher. Speculating that weaker unemployment would mean slower interest-rate increases, investors bid up fixed-income ETFs. The chart below compares the long bond fund TLT, and the corporate bond fund LQD.

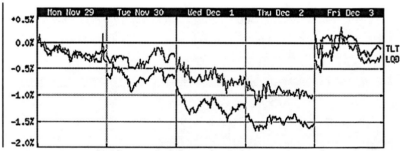

Figure 3

As the chart above shows, the 4-day sell-off in fixed-income ETFs was reversed after the employment report came out pre-market Friday. Bond ETFs immediately opened higher.

Technology ETFs did well. Good news out of Intel (INTC) helped, but technology ETFs were already celebrating mid-week on lower oil. One of the best performers in technology was software (IGV). The chart below compares the returns of IGV with a more general technology fund, IYW.

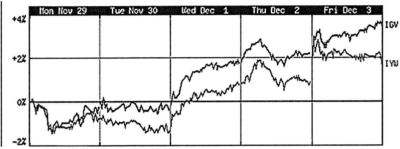

Figure 4

DISCUSSION:

When everything is going well with a portfolio, investing is easy. But the portfolio is long the Bush sector: utilities and oil. Oil collapsed, down $7 in five days. Utilities collapsed. The question is always the same when things go wrong: Is the overall direction of the market changing, or is this a temporary and cyclical move? No security goes up in a straight line. Even the best charts are always a little higher, a little lower; a little higher, a little lower.

Why might this be the time that oil and utilities really start falling? One concern is that oil is increasing in volatility. But over the past year, whenever there has been a sharp, market-direction-changing move in one week, the next week has reversed it. Crude built up a slow and steady increase. A sharp drop is probably a sign to buy. But already long oil, and I don't think I can stomach the losses if oil falls further. I would like to bet against the sharp move downward, just go in there and buy oil and to sell tech. When I have done this in the past, I have been rewarded. But adding to a losing position as the market moves against it is like standing in the face of a lion.

▲ *There is an expression on Wall Street: "There are old traders, and there are bold traders, but there are no old bold traders." The method outlined in this book is to assume that the trend will continue, to always of course be vigilant for signs of a change, but not to predict that change.*

When things get bad: Count. How many weeks has a sector sold off compared with the number of weeks it has gained? What is the size of the loss compared to the size of the gain? Historically what has happened in a sector following a sharp rise or fall? Because the market never goes in a straight line, counting the number of times it has moved in either direction helps to establish the expected incidence of negative weeks in an overall uptrend, and positive weeks in an overall downtrend.

The table shows the number of times over the last six months that the oil sector fund XLE and the utility sector fund IDU have been up, down and the average size of the move as a percentage.

ETF	Up weeks	Down weeks	Unchanged	Average % up	Average % down
XLE	21	5	0	1.7%	3.2%
IDU	17	7	2	1.5%	1.3%

Table 7

On an average down day, XLE moves almost twice as much as it does on an average up day. This week's returns of -4.4% look consistent with how oil has traded. But this week's -3.3% return on IDU is larger than average and, checking back, is the single worst weekly return over the last 6 months.

PORTFOLIO:

I wish I had the guts to buy utilities and oil here on what looks like weakness. But it is important to keep discipline. Cut back slightly on oil and utilities.

ETF	LW	TW	P/L
XLE	5	4	↓4.4%
IDU	4	2	↓3.3%
SPY	2	2	↑0.8%
QQQ	2	2	↑2.3%
TLT	-3	-3	↑0.5%
Portfolio +1.1%			

LW: Last Week | TW: This Week | P/L: Profit/Loss

Week 24: December 6 – December 10

THEME:
- *What Do I Do When the Market Trades Sideways?*

INDICATORS:
- EN, SF, CC

SITUATION:
The market traded sideways as many investors continued waiting and hoping for the seasonal Santa Claus rally. Oil slipped lower, closing at $40.75, even as OPEC agreed to cut supply. Inflation remains an issue: The November producer price index (PPI) came in Friday at 0.5%, more than double what had been expected. The high PPI number ensures that the Fed will raise rates at its meeting next week. Despite the promise of a rate hike, fixed-income advanced.

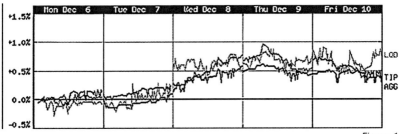

Figure 1

The chart above compares the weekly returns of three bond ETFs AGG, TIP, and LQD. Not only are bond investors not selling, but the inflation-indexed TIP bonds are not outperforming.

One reason fixed-income may be holding up here despite the inflation threat: Broad markets are stagnant to lower. The chart below shows the weekly performance of the QQQ, SPY, and DIA.[1]

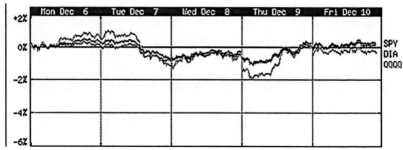

Figure 2

Foreign markets were lower, in part because of a stronger dollar. Late Friday, the dollar traded at 105.2 yen, up from about 102.5 yen a week ago. The dollar also strengthened to 1.32 euros from over 1.34 last week. Asian ETFs were hit hard: Japan (EWJ) down 3.7%, Pacific ex-Japan (EPP), which invests in Australia, Hong Kong, and Singapore markets, down 4.1%. The poorest-performing ETF of the region was the volatile South Korea Index (EWY), down 5.7%. The chart below compares the returns of EWJ, EPP, and EWY.

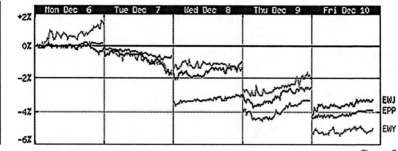

Figure 3

---

1 The Nasdaq 100 Trust formerly traded on the American Stock Exchange under the ticker QQQ. It has now moved to the Nasdaq and trades under the ticker QQQQ.

DISCUSSION:

What to do when the market trades sideways? Nothing. Wait. Although crude continues to tumble, with the broad market stagnant and technology lower, it does not look as if there is any fundamental change ahead yet for oil. Utilities also continue to work. The rest of the market continues to look for direction.

But the short bond position is not trading sideways. It is not working at all. TLT is up again this week, along with the rest of fixed-income. There is too much uncertainty in this market for a short bond position to pay off. I'm exiting the short TLT position here.

PORTFOLIO:

ETF	LW	TW	P/L
XLE	4	4	↓1.2%
IDU	2	2	↓0.2%
SPY	2	2	↑0.1%
QQQQ	2	2	↓0.4%
TLT	-3	Closed	↓1.6%
Portfolio +1.2%			

LW: Last Week | TW: This Week | P/L: Profit/Loss

Week 25: December 13 – December 17

QUESTION:
- *Mergers in the Software Sector – Time to Buy?*

INDICATORS:
- GD, EN

SITUATION:

The big news this week were mergers: Oracle (ORCL) & PeopleSoft (PSFT) finally came together and Symantec (SYMC) & Veritas (VRTS) also announced a merger. But the software ETF (IGV), which might have been expected to benefit from the mergers, is down this week. The chart below compares the weekly returns of IGV as compared with Networking (IGN), and Semiconductors (IGW), and the more general tech-focused IGM.

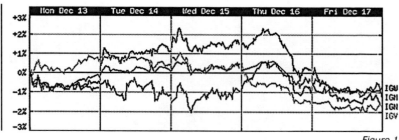

Figure 1

The chart above shows that the software sector fund IGV is

lagging other sectors within technology.

DISCUSSION:

If a string of mergers does not help tech stocks, what will? I take this week's poor performance as confirmation that technology is going nowhere. Leadership has shifted away from technology and shows – at this point at least – no sign of returning. This is a good time to cut back on the cubes trade. Maybe it is time to go back to shorting technology.

PORTFOLIO:

Crude closed the week at $46.28, up over $5 in five trading days. The chart compares XLE and IYE with the benchmark SPY.

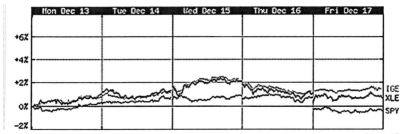

Figure 2

Though oil ETFs have not gone up as much as crude, they continue to outperform the benchmark. This is a good time to add to the position. Utilities also are on fire. Add to the utilities on strength.

ETF	LW	TW	P/L
XLE	4	5	↑1.7%
IDU	2	4	↑2.9%
SPY	2	2	↑0.6%
QQQQ	2	1	↓0.2%
Portfolio +1.2%			

LW: Last Week | TW: This Week | P/L: Profit/Loss

Week 26: December 20 – December 23

QUESTION:
■ *The Pfizer Case: Can One Holding Crash an ETF?*

INDICATORS:
■ TA, FA

SITUATION:
What happens when one of the major holdings of an ETF plunges on negative news? How does this affect the price of the ETF? The recent collapse of Pfizer (PFE) on news that its blockbuster painkiller Celebrex causes cardiovascular risk in patients taking high doses provides a test case.

Last Friday, December 17, Pfizer lost 11.2% of its value and traded to 7-year lows. The largest pharmaceutical company in the world, Pfizer was a major component of every health care index. It was the top holding for three of the five healthcare ETFs. It represented 13.98% of the total assets in iShares Dow Jones US Healthcare (IYH), 14.52% of Vanguard Health Care VIPERs (VHT), and 16.26% in Health Care Select Sector SPDR (XLV). Pfizer is somewhat of an exception. In general, it is unusual to have as much as 15% of an ETF's holdings represented by a single company. The chart below shows the performance of VHT and XLV before and after the Pfizer announcement.

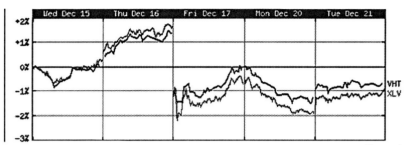

Figure 1

By Tuesday, December 21, just three trading days after Pfizer's announcement, VHT stabilized to trade at a level it attained Wednesday before Pfizer's announcement. XLV also traded close to its Wednesday low.

DISCUSSION:

ETFs are about indexing-- buying a basket of representative stocks when investing in an asset class. Usually, buying the ETF means buying an index and avoids pitfalls associated with individual stock picking. But this does not mean that ETFs are invulnerable to bad news in specific stocks.

Despite the dramatic drop in Pfizer, VHT lost 1.5%, IYH 1.6%, and XLV 1.7%. While this is hardly a stellar performance for these ETFs, and by far the most dramatic, as the chart below comparing the returns of VHT and XLV shows, this was by no means fatal news for these ETFs. Most importantly for these ETFs, the market judged Pfizer's woes to be company-specific. The fall-out at Pfizer did not spread to other companies in the MSCI US Investable Market Health Care Index, which VHT tracks.

This case shows that although top holding Pfizer suffered a major setback, the overall impact on the performance of these three healthcare ETFs with major Pfizer holdings was limited.

PORTFOLIO:

Financials also moved to new 52-week highs, resuming a stalled post-election break-out rally. The chart below shows how financial ETFs resumed their post-election rally this week by comparing the 6-month return of XLF, IYG, and IYF with the benchmark DIA.

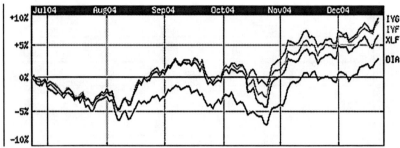

Figure 2

This week, defensive sectors again were popular with investors, as they have been all year. Utilities powered to new 52-week highs. The two charts below show the 5-day and 52-week return on two utility ETFs, IDU and XLU. The divergence of IDU and XLU Thursday reflects a dividend payment.

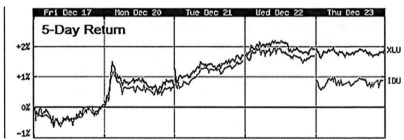

Figure 3

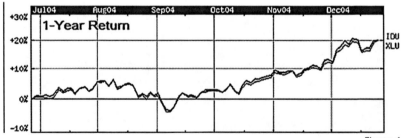

Figure 4

I like the long positions in oil and utilities. It seems like they can do no wrong. I want to add utilities back in again here on these new highs. Also the DIA. And I want to add XLF, the financial ETF, because it too is hitting new highs here. I'm going to hold on to the

QQQQ short.

ETF	LW	TW	P/L
XLE	5	5	↑1.3%
IDU	4	5	↑1.6%
SPY	2	2	↑1.3
QQQQ	1	1	↑0.7%
DIA	0	1	
XLF	0	1	
Portfolio +1.3%			

LW: Last Week | TW: This Week | P/L: Profit/Loss

Week 27: December 27 – December 31

QUESTION:
- *Are Trends in 2004 Important for 2005?*

INDICATORS:
- TA, CC, EN, FA

SITUATION:
First the weekly round-up: ETFs ended negative, despite good news and technical strength. The benchmark SPY traded its 2004 high on Friday, the last day of the year, but sold off sharply, closing on its weekly low. The DIA also made its way to multi-year highs earlier in the week, before crashing and ending Friday on its weekly low. The chart below compares the SPY and the DIA.

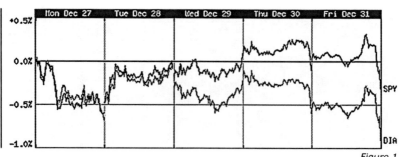

Figure 1

Broad market ETFs initially were helped by an impressive

showing in the December consumer confidence number, which trumped expectations Tuesday, coming out at 102.3, more than 8 points higher than expected. Oil ended the week down slightly, at $43.45 a barrel.

A weak dollar helped many foreign ETFs to outperform domestic funds again this week. The chart below shows this by comparing three popular foreign ETFs, the Japan EWJ, Brazil EWZ, and Canada EWC with the benchmark SPY.

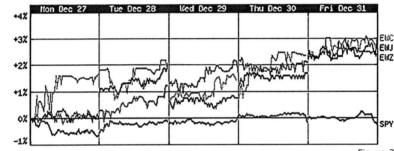

Figure 2

DISCUSSION:

The main trends of 2004 were higher oil and commodity prices, rising commercial property values, a listless equity market, higher deficit and a lower dollar, boosting foreign-denominated assets. The table below shows the 10 best-performing ETFs for 2004:

NAME	TICKER	2004 RETURN
iShares MSCI Austria Index	EWO	69.9%
iShares MSCI Mexico Index	EWW	50.1%
iShares MSCI South Africa Index	EZA	42.5%
iShares MSCI Belgium Index	EWK	42.3%
iShares MSCI S&P Latin America Index	ILF	37.9%
iShares MSCI Sweden Index	EWD	35.1%
iShares Cohen & Steers Realty Majors	ICF	33.4%
Energy Select Sector SPDR	XLE	32.5%
streetTRACKS Wilshire REIT Fund	RWR	31.5%
iShares Dow Jones US Real Estate	IYR	29.2%

Table 8

The worst-performing ETFs for 2004 were in technology and

health care. Reduced corporate spending and disappointing profits hurt tech. Health care lost its post-bubble momentum.

The table shows the 10 worst-performing ETFs for 2004:

NAME	TICKER	2004 RETURN
iShares Goldman Sachs Semiconductor	IGW	-15.1%
iShares Dow Jones US Technology	IYW	0.2%
iShares Lehman 1-3 Year Treasury Bond	SHY	0.6%
Healthcare Select Sector SPDR	XLV	0.8%
iShares Goldman Sachs Technology	IGM	2.1%
iShares S&P Global Technology Sector	IXN	2.4%
iShares Nasdaq Biotechnology	IBB	3.2%
iShares Dow Jones US Healthcare	IYH	3.2%
iShares S&P Global Healthcare Sector	IXJ	4.7%
iShares MSCI Taiwan Index	EWT	5.0%

Table 9

The only negative ETF and overall worst performer in 2004 is IGW, down 15.1%. Chip companies are often seen as an indicator of the health of technology and the overall market.

This is a good time of year to look back, to look ahead, and to position for the year to come. Will the rally continue in 2005? If so on, what themes? Will it be a repeat of 2004: oil, utilities, real estate, and a weakening dollar? The poor showing of IGW in 2004 is not a good sign for the market in 2005. Will 2005 otherwise resemble 2004? With Bush re-elected, I think we can expect more of the same. The first weeks and months of the year are critical, as they often give an indication of what themes will continue and what will not.

PORTFOLIO:

Movement in the last week sent a mixed signal. As the chart in figure 1 above shows, on the one hand, the QQQQ and DIA traded to new highs mid-week. This suggests that they are going higher and that this would be a good time to buy. On the other hand, these key ETFs were unable to sustain these highs and in fact closed out the week and the last couple of hours of the year with sharp losses.

Ultimately, as the tablebelow shows, none of the ETFs in the portfolio have moved very far from where they were at last week's close. No change.

ETF	LW	TW	P/L
XLE	5	5	↓0.5%
IDU	5	5	↓0.3%
SPY	2	2	0%
QQQQ	2	2	↑0.4%
DIA	1	1	↓0.8%
XLF	1	1	↓0.3%
Portfolio -0.2%			

LW: Last Week | TW: This Week | P/L: Profit/Loss

Part III, Q1 2005

~

Week 28 - Week 40

Week 28: January 3 – January 7, 2005

QUESTION:
■ *What Do We Do When Everything Goes Wrong?*

INDICATORS:
■ GD, CC

SITUATION:
The first week of 2005 was a rout. The market was hurt by tax selling, concern about higher inflation, and speculation about a more aggressive Fed. Once the selling began, no catalyst emerged to change the market's direction. ETFs fell lower – and lower. The chart below shows the returns of the QQQQ, DIA, and SPY.

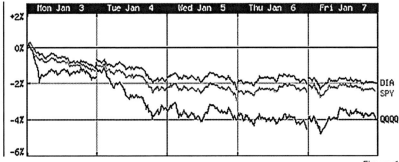

Figure 1

Although on Thursday and Friday, the broad market evened out, positive market-moving news remained scarce. Investors looked to the December employment report for relief and support, but the numbers surprised few: Unemployment came in line.

Other ETFs fared worse. Technology ETFs did not have a good 2004, but in November and December they mostly redeemed themselves, to post slight gains by year-end. Those slight gains in 2004 were reversed in the first three days of 2005. The chart below compares the returns of software (IGV), networking (IGN), and the semiconductors (IGW) with the tech sector benchmark IGM.

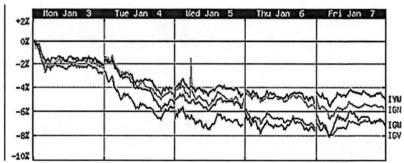

Figure 2

Often thought of as defensive and counter-cyclical, REITs nonetheless disappointed, hurt by the Fed minutes released Tuesday. Higher rates increase the cost of borrowing and can negatively impact highly leveraged REITs. The chart below shows how REITs ICF and RWR traded with the benchmark SPY, before cascading lower Wednesday.

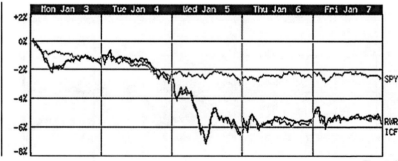

Figure 3

The fear of aggressive rate hikes gripped investors who took REIT ETFs to their steepest single-day drop since April of last year: a fall of 7% in three days. The sharp move in REITs lagged the broad market.

Meanwhile, while all this chaos and market slaughter is afoot, gold is falling, but bonds are calm, unmoving, like the eye in the storm. The chart below compares the week's return of Gold Shares (GLD) and the returns of two fixed-income ETFs, the Aggregate Bond AGG and the Corporate Bond LQD.

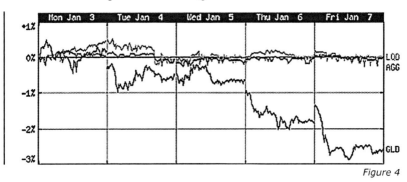

Figure 4

DISCUSSION:

What to do when things go wrong? Get out and cut back! Welcome to 2005. Only the shorts made money. When things go badly, I get out of things that are not working and get into things that are working. What is working in the portfolio? Nothing. So I'll get out and cut the size of everything. Oil and utilities are defensive positions and should be relatively strong even if the market continues to fall. But I need to stick to the strategy and cut the size of these positions in case the sell-off continues next week.

The SPY, QQQQ, DIA, and XLF positions are easier than oil to jettison. I'm nervous about holding the long financials position anyway. Last week, they did not follow through on their highs of two weeks ago. This week, it was the same. I'll take the loss. It's time also to get out of the broad market DIA, which is falling.

Fixed-income stagnation here is a bit of a mystery. The fact that fixed-income did not move higher, despite a falling market and strengthening dollar, suggests perhaps that the market is selling more on momentum than on uncertainty.

The fall in gold is interesting. It is selling right along with tech and the broad markets. It looks as if people were buying gold because of the dollar depreciation. With the dollar on a huge rally trading to 1.305 to the euro late Friday, investors figured they no longer needed to hold gold.

What about trends for the New Year? Two obvious differences; dollar strength and lower REITs. The fall in REITs seems particularly notable. Sure, technology sold very heavily, but technology has been selling. REITs are supposed to be defensive and have low market correlation. REIT ETFs were among the best performers in 2004 – buy in and forget about it. But it is not looking that way at all this year. Falling right with the market but faster and farther, they are among the worst performers and do not look like low-beta funds. It looks very bad to be long REITs here.

Why are REITs selling? Fear of rising rates and profit-taking? But what is new about that? Why now? Many market watchers have been calling REITs fully valued since before the Fed started raising rates last year in June. The chart below shows this by comparing IYR with the benchmark SPY over a 5-year period. The recent fall in REITs does not even appear on a 5-year chart.

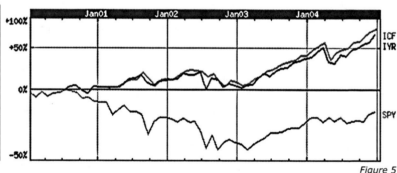

Figure 5

There are reasons to believe that the fall in REITs will not continue. The first worry for any short is that when REITs move lower, their yields (the dividends they pay) improve. With yields on bonds so low and Treasuries paying so little, many investors are looking to REITs for dividends. The increase in yields encourages these investors. The second reason is that if the broad market continues to struggle in 2005, as it now looks as if it might, REITs may still be a good place to take refuge. Third, when the market is weak, many investors believe that the Federal Reserve is less likely to raise interest rates, and this should also help REITs. If future rate increases are truly moderate, REIT ETFs could come out winners yet again.

On the other hand, ICF closed the week on its lows: 7% is a sharp drop. REITs have been strong since the beginning of 2003, so maybe this is the beginning of a trend lower. One reason REITs have done well over the last 52 weeks is that the underlying real estate held in REIT portfolios is highly leveraged. REITs therefore tend to be especially sensitive to interest rates. When ICF was started in early 2001, the target federal funds rate was 5.5%. Greenspan cut interest rates very forcefully in 2001 and did not begin raising rates again until just six months ago, in June 2004. I'm going to sell ICF here.

The biggest opportunity may be in technology. If in the first three days of the year, technology undid all of last year's gains, what will keep tech from going lower next week, now that anyone who bought in all of 2004 is facing a loss? I'm selling IGM.

PORTFOLIO:

ETF	LW	TW	P/L
XLE	5	3	↓3.8%
IDU	5	2	↓3.1%
SPY	2	Closed	↓2.3%
QQQQ	2	Closed	↓3.4%
XLF	1	Closed	↓1.5%
DIA	1	Closed	↓1.8%
ICF	0	-1	
IGM	0	-2	
Portfolio -2.9%			

LW: Last Week | TW: This Week | P/L: Profit/Loss

Week 29: January 10 – January 14

QUESTION:
- *Defense or Offense?*

INDICATORS:
- EN, GD, CC, TA

SITUATION:
A key measure of inflation, the December PPI (Producer Price Index), came in much better than expected. But the market finished this second week of 2005 in the red. At $48.38 a barrel, oil is the problem. Bad news for the market but good news for energy funds. The chart shows energy XLE and IGE and the benchmark SPY.

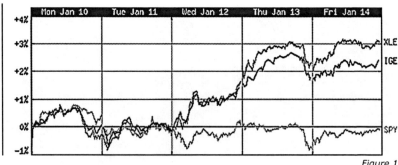

Figure 1

Investors also bought fixed-income products, which are moving higher this year, despite the inflation threat and the promise of rate hikes from the Fed. Fixed-income ETFs were helped this week by continuing strength in the dollar and positive comments on inflation from European Central Bank President Jean-Claude Trichet. The chart below compares the long bond TLT, the mid-term bond fund IEF, and the corporate bond fund LQD.

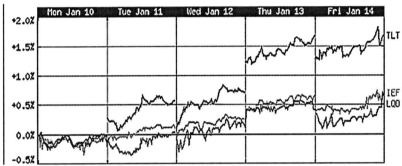

Figure 2

The chart above shows TLT outperforming other bond funds. Given broad market weakness and dollar strength, this chart shows results that I would have thought to have materialized last week. The utility sector also had a good week. Utilities pay high yields and have been performing well in what many investors believe is a regulatory environment friendly to the industry. The chart below shows this week's returns of IDU and XLU.

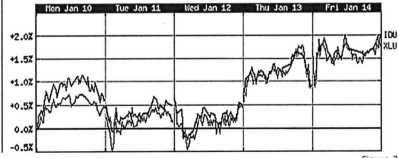

Figure 3

One of the worst-performing sectors this week was telecom. The chart below compares the returns of the telecom fund IXP with health care IYH, technology IGM, and the financial sector fund

IYF.

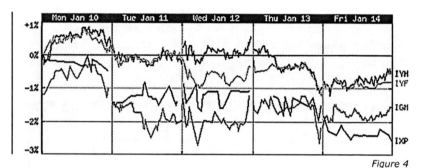

Figure 4

DISCUSSION:

Sometimes it is a no-brainer. Defensive sectors improve. Tech and telecom fall apart. What is defensive? Fixed income, oil, utilities. I want to increase these positions. What looks weak? Tech and telecom. I want to increase the tech shorts. Buy the highs, sell the lows.

REITs are also defensive, too, but I am short here. I'd really like to get my REIT position right for a change. There is no bounce in REITs yet, so this is at least not negative for the portfolio. But there is little follow-through on last week's selling. This suggests that investors are still looking for defensive plays. With the dividend-paying utlities and bonds, investors are chasing yields in the REIT sector and there is risk here for being short.

But there are also reasons for staying short REITs. Low rates have provided immense benefit to REITs, not only because of their leveraged portfolios, but also because REITs are required by law to pay out at least 90% of their taxable income in the form of dividends. These dividends mean that REITs pay healthy yields, which are especially valued at times when interest rates are falling and yields on other dividend-producing instruments are falling. If rates continue to rise in 2005, REIT yields will have increasing competition from other income products. In other words, with rates going up, this advantage is disappearing.

So although REITs are defensive and most defensive sector funds are improving, I want to stay short REITs at this level.

PORTFOLIO:

ETF	LW	TW	P/L
XLE	3	6	↑4.3%
IDU	2	4	↑1.8%
QQQQ	0	-1	↑0.3%
ICF	-1	-1	↑1.1%
IGM	-2	-4	↑0.4%
Portfolio +1.8%			

LW: Last Week | TW: This Week | P/L: Profit/Loss

Week 30: January 18 – January 21

QUESTION:
- *Will Technology in 2005 Be a Repeat of 2004?*

INDICATORS:
- GD, PD, BY, FA, TA

SITUATION:
Political rather than economic events took center-stage this week. The markets, like the nation itself, was focused on the Senate review of Secretary of State Condoleeza Rice and the second inauguration of the freshly elected President George W. Bush.

Sometimes political news can move markets. The November and December rally last fall started when President Bush won the election. But there was little cheering this week. Traders and investors sold the inaugural party, sending the broad market lower for the third straight week, to levels near or even below their pre-election rally.

The market closed out this short and difficult week with big losses. Economic news was not bad: December CPI came in Wednesday ahead of expectations. Housing starts in December were also stronger than expected. Crude ended the week unchanged. The chart below shows the 1-week return of the Nasdaq 100 Trust (QQQQ), the Standard and Poor's Depositary Receipts (SPY), and the Diamonds Trust (DIA).

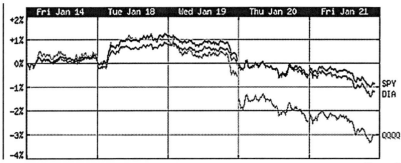

Figure 1

The weakest funds were in technology and biotech. The chart below compares telecom fund IYZ, biotech fund IBB and IGM.

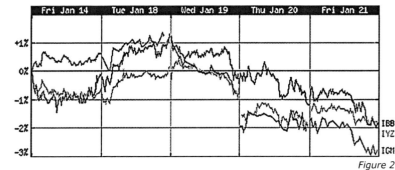

Figure 2

As tech and biotech sold, fixed-income ETFs rallied. The chart shows this week's returns of the long bond TLT, with the corporate bond LQD, near-term bond SHY, and the aggregate bond AGG.

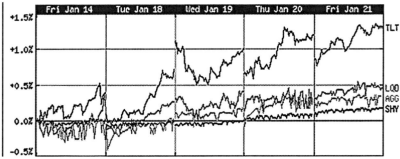

Figure 3

Bonds were helped by overall market anxiety, poor performance in the equity markets, and remarks from the Federal Reserve that inflation was under control. TLT closed near its 52-week high, set in early March of last year, before the Fed started raising rates. Clearly, the Fed's rate hikes are having little impact on bond yields.

DISCUSSION:

In some ways, this week's trading followed a pattern set last year: REITs and energy were leaders; utilities outperformed; fixed-income was stronger than expected. Technology lower. The chart compares ETFs from the strongest sectors, ICF, XLE, and IDU.

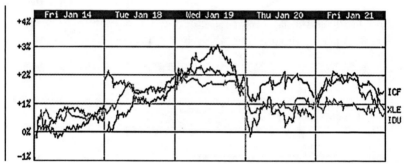

Figure 4

Is this a further indication that 2005 will repeat 2004? Last year, REITs did well because the broad market did badly. ICF's beta is 0.34, among the lowest betas of any ETF. This low beta suggests that historically, REITs have shown a very low correlation to the broad market. The market doesn't look as if it is improving this year, and this should once again prove positive for REITs. I'll reverse the short position and open up a long REIT position.

What about technology? The contrarian view here would be that 2005 will not be like 2004, and technology will return to leadership. If this is to be, then semiconductors are the sector to own. IGW was the only ETF that lost money in 2004, and it lost more than 15%. This outsized negative return means that it radically underperformed all other sectors, including its peers within the technology sector. Just a week into 2005, IGW is already down another 10%. It has fallen faster and farther than virtually any other sector and is now trading at a level below the pre-election rally in November of last year.

There is also the fundamental story: 2004 was a good year for chip sales, with revenues estimated at around $200 billion for the industry, but semiconductor companies overbuilt and overproduced. The problem seems to be that expectations of high demand in 2004 led to a 50% surge in capital spending in the industry. By mid-year, supply significantly outpaced demand, and chip companies became saddled with inventory. In the first half of 2005, the industry faces some of the same trouble as last year: reduced capital spending, lower utilization, and pricing pressure.

Growth estimates for the industry in 2005 vary, but most remain negative or in the low single digits. The story is that businesses are reluctant to spend money on software or hardware. In 2004, the hot iPod MP3 player from Apple Computer (AAPL) motivated consumers to spend. Increasing consumer demand for electronic devices will be some help clearing semiconductor inventory in 2005. But there is currently no "must-have" technology equivalent to the iPod for business customers, and there is none on the immediate horizon. "Longhorn," the next-generation operating system from Microsoft (MSFT), anticipated as a driver for corporate hardware spending, will not be available before 2006.

Technically, IGW does not look good. It closed last week at 50.50, below its 50-day and 200-day EMA. This technical situation is important because both short- and long-term traders watch the 50-day and 200-day EMA to assess price direction. IGW broke through support at about the 53 level. Technically, IGW does not have support until the 48 level, and then again at 45. The one-year price chart shows IGW falling below its 50-day and 200-day EMA.

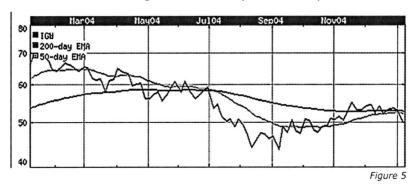

Figure 5

Crossing below both the 50-day and 200-day EMA, IGW looks

very bad. IGW's top five holdings tell a remarkably similar story. With the exception of Motorola (MOT), which had new management installed at the end of 2003 and is more diversified than the other IGW holdings – Intel Corp (INTC), STMicroelectronics (STM), Applied Materials Inc. (AMAT), and Texas Instruments (TXN), all show price performance very similar to IGW and to each other. The 1-year chart below compares the returns of IGW's top five holdings: MOT, STM, INTC, TXN, AMAT.

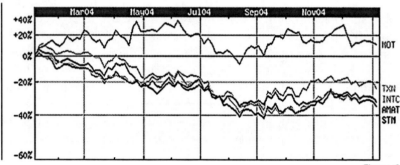

Figure 6

On the other hand, IGW never looked better than it did exactly a year ago, in January of 2004. IGW's 75% return in 2003 put it among the top-performing ETFs for that year. Forecasts for semiconductors were bright. Chip companies were expanding and increasing capacity. But the contrarian bet paid off: January of last year was a time to sell IGW. The situation is the reverse now. The semiconductors are shunned, the forecast is gloomy. Maybe it is time to be contrarian again.

But, the play is not without risk. IGW's beta is 2.35, currently the second-highest of any ETF, so even small market swings can produce big moves. Still, as bad as last year was, IGW was up almost 20% in the last two months.

Because of this volatility, it is vital to time IGW well. Technically speaking, this is not yet the place to begin buying IGW. The place to buy (to reference Figure 5 above) is either as IGW crosses its EMA around the 53 level or falls to support at the 45 level. After last year's rout, I think there is a good chance for semiconductors this year. But I want to wait to buy IGW at either 45 or 53, whichever happens first.

⚠ *It is not easy to catch the bottom or sell the top. An investor should usually be content to capture just part of the move. Even catching a quarter or an eighth of the move can provide a decent living.*

PORTFOLIO:

Right now it looks like a repeat. Technology is collapsing, ending the week on its lows. It looks like ICF is trading with other defensive sectors. I want to reverse the REIT position and add to the short positions in tech here.

ETF	LW	TW	P/L
XLE	6	6	↓0.1%
IDU	3	3	↓0.1%
QQQQ	-1	-2	↑3.6
ICF	-1	1	↓1.0%
IGM	-4	-5	↑3.4%
Portfolio +0.8%			

LW: Last Week | TW: This Week | P/L: Profit/Loss

Week 31: January 24 – January 28

THEME:
- *What Is an "Untradable Sector?"*

INDICATORS:
- **GD**

SITUATION:

The broad market ended a hectic week little changed. Economic numbers disappointed: December durable goods came in below expectations. Fourth-quarter GDP also disappointed, coming in below the 3.5% expected, at 3.1%. The lack of a significant updraft this week all but guaranteed that January of 2005 would be the worst January for the broad market in over 20 years. The chart shows QQQQ, IGN, IGW and the software sector ETF IGV.

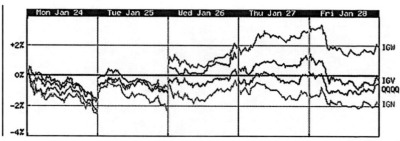

Figure 1

The week started Monday with a technology sell-off, which pulled the QQQQ to its worst levels since before the election in October of last year. As the above chart shows, the networking sector (IGN) was the weakest in the technology group this week, while the semiconductor (IGW) was the strongest.

Fixed-income looked for direction this week, though there was some volatility, particularly in the long bond fund TLT. Fixed-income jumped higher Friday on the disappointing fourth-quarter GDP numbers. The chart below compares the returns of a short-term, mid-term, and long-term bond funds SHY, IEF and TLT respectively.

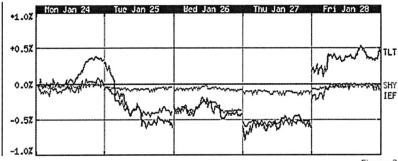

Figure 2

Among the biggest losers this week: REITs. REITs ended the week down about 4% and on track to end their worst month since April of last year. REITs have historically had among the lowest correlation of any sector to the broad market. But this isn't helping REITs now. The chart below compares this week's returns of two REIT ETFs, IYR and ICF, with the benchmark SPY.

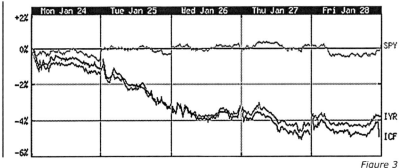

Figure 3

DISCUSSION:

A sector is "untradable" when it does not respond well to the trading strategy being followed. The basic strategy here involves buying on strength and selling on weakness, based on a weekly metric. Often REITs seem to lack follow-through from one week to the next. Despite impressive gains, I am not comfortable owning REITs in an environment of increasing rates in the first place. I should have just been patient and held on to the REIT short.

⚠ *It is always very dangerous and arrogant for a trader to reverse a position (move from long to short or short to long on the same sector as opposed to merely getting out of a position), when it doesn't work. This is trading as if he could time the market perfectly. This can be an acceptable strategy in rare instances, when a trader is ahead in a sector and reading it well. But this is a terrible idea and deadly practice if the position is a losing one, as decisions under stress tend to be emotional and desperate. Sometimes when a trade starts off badly, as the REIT trade did last year, a trader becomes traumatized and tortured by the position, and then the best thing is just to stay out.*

Now I will get out of ICF and stay away. I need to take the view that REITs are like Latin American ETFs – very interesting, with massive swings, providing big opportunities when they can be timed, but as this is hard to do, they are distracting. Very interesting, but ultimately untradable.

PORTFOLIO:

As the chart in Figure 1 above shows, overall this was a volatile week, with huge after-hours movement. Broad market ETFs dropped significantly on Monday, opened almost a full percentage point higher on Tuesday morning, fell Tuesday during the day and once again opened up much higher on Wednesday. This kind of market poses opportunities for contrarian traders taking advantage of fast-moving markets that overshoot. But I am trading with the trend. Choppy markets suggest that the trend has stalled, or is in the process of consolidation. A choppy market in general is not a good time to add to a portfolio. Except for getting out of the rotten ICF position, no change.

ETF	LW	TW	P/L
XLE	6	6	↑2.0%
IDU	3	3	↑1.8%
QQQQ	-2	-2	↑0.3%
ICF	1	Closed	↓3.8%
IGM	-5	-5	↑0.2%
Portfolio +1.7%			

LW: Last Week | TW: This Week | P/L: Profit/Loss

Week 32: January 31 – February 4

QUESTION:
- *What Do OPEC's Announcements Tell Us?*

INDICATORS:
- GD, EN, BY

SITUATION:
This was a good week for the market. The week began Monday with news that the Iraqi elections appeared to be successful. That was a surprise, and the markets celebrated. On Tuesday, the Federal Reserve hiked interest rates, which often has a negative impact on equities. But this time the market was reassured: just the expected 25 basis points.

The only news that really threatened broad market ETFs was the January employment report, out Friday. Nonfarm payrolls for January disappointed, coming in substantially under the 200,000 expected at 146,000. But here too traders chose to focus on positive implications of the job report: Higher unemployment means less chance that the Fed will accelerate rate hikes and also mild inflation. In a positive sign for the bulls, small-cap and mid-cap ETFs mostly beat out larger caps. The chart below compares three of the new Vanguard funds: the large-cap Vanguard Total Stock Market VIPERs (VTI), the mid-cap Vanguard Extended Market Index VIPERs (VXF), and the Vanguard Small-Cap VIPERs (VB).

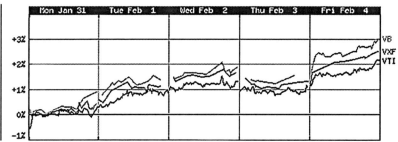

Figure 1

As the chart above shows, the market performed well.

But outperforming the broad market were the energy ETFs. One of the top asset groups in 2004, energy ETFs have come running out of the gate in 2005. This week, they put on their best show this year, up between 5% and 6% in five trading days. Energy ETFs were helped by OPEC's acknowledgment over the weekend that the price target for a basket of crudes, in place since 2000, would be suspended (and later implicitly revised upward). The chart below shows the weekly returns of three energy ETFs, Energy Select Sector SPDR (XLE), iShares S&P Global Energy Sector (IXC), and iShares Goldman Sachs Natural Resources (IGE).

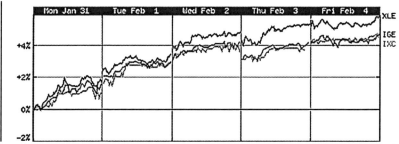

Figure 2

One of the week's surprises was the move in the long bond TLT. TLT spent most of the week unchanged to lower, but jumped a stunning 1.5% Friday to a new 52-week high on the weak unemployment report. The move in TLT reflects a 1-month low on yield of the benchmark 10-year Treasury note. The move in TLT is all the more remarkable considering that other fixed-income ETFs reacted less dramatically, including the mid-term Treasury bond ETF, iShares Lehman 7- to 10-year Treasury (IEF). The chart below compares TLT, IEF, and the iShares Lehman Aggregate Bond (AGG).

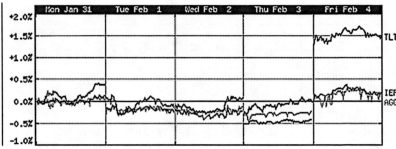

Figure 3

Utility ETFs are on the move. The bull market in utilities ETFs started in the beginning of 2003, but unlike ETFs representing the broad market, continued strongly last year. They are up 6% in two weeks, and moved up 3% this week to new 3-year highs. The chart below shows this week's returns of two utilities, XLU and IDU.

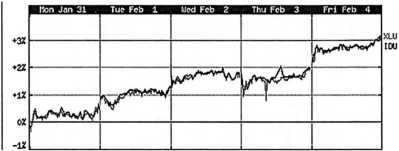

Figure 4

DISCUSSION:

This business of OPEC suspending its 5-year old oil price band mechanism, formerly $22 to $28 for its reference basket of seven crudes, is really big news and not getting the attention it deserves. This may seem like an insignificant move by OPEC, given that the price of the reference basket was $42 when they made the announcement and has not traded within the proposed band for over a year. But the OPEC move is not trivial and looks as if it marks a watershed moment. The decision underscores what appears to be a new reality: Oil prices are not expected to make a sustained retreat any time in the foreseeable future.

This does not mean that there are no seasonal demand trends. In spring, for example, as the demand for heating oil falls, oil prices are historically weak. Seasonal oil price fluctuations are expected to continue. They may fall next week or the week after. However, just six months ago, popular sentiment was that oil prices were high, gasoline prices were high, but that this was likely temporary. After all, energy prices fluctuate. This sentiment is gone.

OPEC's decision to suspend the price-band mechanism points to something different: a structural change in oil demand and higher prices. The chart below shows the price of a barrel of West Texas Intermediate over a 52-week period.

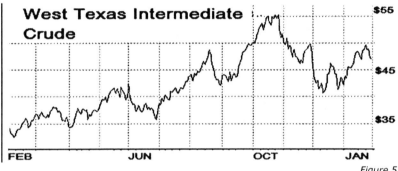

Figure 5

The chart below compares the 52-week return of two oil service ETFs, XLE and IGE.

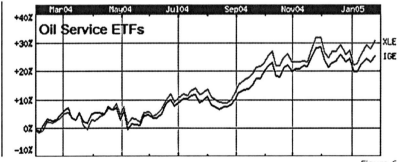

Figure 6

A comparison of the two charts above shows that oil the commodity has been more volatile than ETFs holding the companies that service it. As the price of crude jumped from about

$35 a barrel in June to about $55 a barrel in October (an appreciation of over 50% in just four months), oil ETFs also improved. But the ETFs' gain was only about half as much, or 25%. When the price of crude fell sharply in November and December, energy ETFs did not retreat. Now oil is trading higher again, though off earlier peaks, and energy ETFs are flirting with new highs.

OPEC members have historically thought in terms of a supply-and-demand curve. Remove supply, oil costs more. But if oil costs too much more, there is a risk that higher oil will send the world into an economic slump, ultimately hurting world demand for oil. At very high prices, alternative sources of energy become viable, further hurting demand. For this reason as well as others, such as the importance of cheap oil to fighting the Cold War, OPEC, and particularly Saudi Arabia, the world's largest exporter and low-cost producer, have historically agreed to seek to maintain a price level favorable to economic development. But OPEC's formal decision to abandon its price band affirms what the market is already indicating: The curve is shifted higher.

Last year, rising oil prices were driven by changes in both supply and demand. On the supply side, political trouble in Venezuela restricted supply. Iraq also provided less oil than had been anticipated. Hurricanes in the Gulf knocked oil rigs off-line. On the demand side, the most important enduring factors were the increasingly thirsty China and India fueling their economic boom.

Like OPEC, investors are looking at the numbers: Last year, the global economy grew at about 5%. For its part, the U.S. economy grew an astonishing 4%, despite some of the highest oil prices seen in years. The conclusion seems to be that high oil prices are not the threat to the global economy that they once were. Though OPEC ministers from different nations vary in their opinions of an ideal price level per barrel, many say that oil at $50 a barrel or higher is not a threat to the world economy.

The risk here is that OPEC has it wrong. Global expansion *and* higher oil prices may not ultimately be sustainable. Last year, a number of factors arguably contributed to global expansion and increased demand for oil, including low interest rates globally, a weak dollar, and massive deficit spending by the United States. Although higher oil and gasoline prices are no longer nightly news, and the economic costs of higher oil have not gone away, OPEC is saying that it is comfortable with higher oil. This is positive for oil ETFs. The way to take advantage here is to buy more for the portfolio.

But the rest of the market looks vulnerable. As the chart in Figure 1 above shows, traders seem to have ignored the bad news, sending up prices for the broad market, despite the poor employment numbers and what looks like higher oil ahead. The resilience of defensive investments like bonds shown in Figure 3, and utilities, shown in Figure 4 above make this broad market rally look temporary. Still, I will keep discipline and trim the short positions in tech that suffered losses this week.

PORTFOLIO:

As the chart in Figure 4 above shows, utility ETFs are on fire. Utilities continue to look like an easy trade. Steady, low beta, no big up-and-down swings, ending the week at an all-time high. This is the kind of steady chart a trader likes to see. It flashes green, it flashes buy. Buy this all day long.

Overall long defense and short offense. This remains the trade. But the way this market is moving, the easiest trades are on the long side: oil and utilities. Why even look at anything else? I should have stuck to my strategy and added on strength last week. Based on the OPEC news, I'm going to step and double the oil position. I need to cut back on the losing tech short.

ETF	LW	TW	P/L
XLE	6	12	↑5.2%
IDU	3	6	↑3.7%
QQQQ	-2	-1	↓2.2%
IGM	-5	-3	↓2.0%
Portfolio +1.4%			

LW: Last Week | TW: This Week | P/L: Profit/Loss

Week 33: February 7 – February 11

QUESTION:
- *Why Are Long-term Bonds Outperforming Short-term Bonds?*

INDICATORS:
- **GD, EN, BY**

SITUATION:
The markets were relatively kind to investors this week, despite Thursday's announcement of a $56.4 billion U.S. trade deficit in December, which pushed 2004 to an all-time record yearly deficit of $617.7 billion. The chart below shows the return of QQQQ with DIA and SPY.

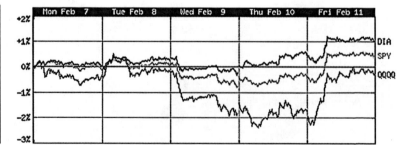

Figure 1

But as the sell-off in the chart above shows, there is still fear in the markets.

Fear has been good for the bond world, where the long bond fund TLT continues to outperform the intermediate-term fund IEF and near-term fund SHY. The six-month chart below shows the performance of TLT, IEF, and SHY.

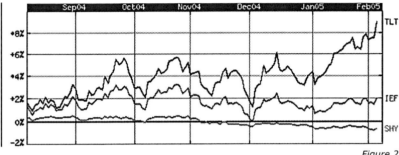

Figure 2

But all three ETFs hold long-term highly rated government Treasury bonds. Why has TLT outperformed so dramatically?

DISCUSSION:

The answer concerns the maturity of the Treasury bonds held in the ETF. The maturity of a bond is the length of time before the debt becomes due for payment. In this case, the U.S. government that issued the bond pays the principal back when the bond matures. The chart above shows that in the last six months, investors have favored ETFs holding bonds with longer maturities like TLT, rather than bond ETFs with shorter maturities like IEF and SHY. The average maturity of the bonds held in TLT is about 23 years, compared to a maturity of about 8 years for IEF, and less than 2 years for SHY.

Why would an investor prefer to have his principal returned later rather than earlier? The answer of course is that, all things being equal, an investor would prefer to have his principal returned earlier, but if he is paid more interest – a higher yield – for the return of the principal later, this option becomes much more attractive. This is the situation currently: ETFs holding bonds with longer-dated maturities like TLT pay a higher yield than ETFs with shorter-dated maturities.

The chart below shows exactly how much more interest bond investors receive for owning bonds with longer-dated maturities.

The chart is called a yield curve. The curve compares the yields on government bonds of different maturities. Yield is represented on the *y*-axis and maturity on the *x*-axis.

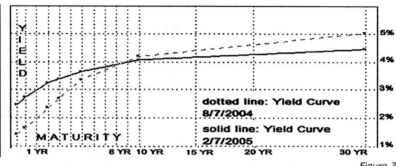

Figure 3

The chart above, in fact, shows two yield curves. Today's yield curve is represented with a solid line. The historical yield curve as it looked six months ago is represented with a dotted line. As the chart shows, over the last six months, as the Federal Reserve has raised rates, yields on fixed income have changed, but not uniformly. In other words, the entire yield curve has not simply shifted upward.

Here is what has happened: yields at the short end of the curve, on the 1-year, 2-year, and 3-year Treasurys, have gone up. Yields in the middle of the curve – the benchmark 10-year Treasury – have held steady, and yields at the long end of the curve have fallen. In short, over the last six months, the yield curve has flattened.

Because price moves in the opposite direction of yield, as the yield on short-term bonds has increased, as shown in Figure 2 above, the price of these bonds has fallen. SHY, which holds bonds with near-term maturities, has drifted lower. The converse is true at the long end of the curve. As yields on the longer-term bonds have fallen, their price has increased. TLT, which holds these longer-term bonds, has increased in price.

At the short end of the curve, the difference between the solid black line and the dotted blue is about 75 basis points 0.75% over the last six months. At the long end of yield curve the difference is about 50 basis points (or 0.5%) in the last six months. The middle of the curve is relatively unchanged. The shift in the yield curve is reflected in the price movement of SHY, IEF, and TLT.

PORTFOLIO:

Although I've sworn off trading EWZ, the chart continues to look phenomenal. Specific country and regional funds are typically very volatile, but strong industrial production, lower-than-expected inflation, continuing higher oil, and what appears to be genuine economic reform mean big returns for investors. The chart below compares Mexico (EWW), Brazil (EWZ), and Latin America (ILF).

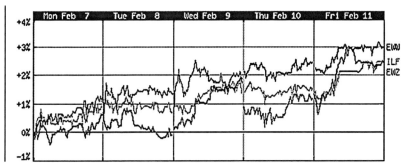

Figure 4

I'll open a position in EWZ on these highs. I added significantly to the oil position last week. I'll pause before adding more.

ETF	LW	TW	P/L
XLE	12	12	↑3.5%
IDU	6	6	0%
QQQQ	-1	-1	↑0.1%
IGM	-3	-3	↓0.2%
EWZ		1	
Portfolio +1.4%			

LW: Last Week | TW: This Week | P/L: Profit/Loss

Week 34: February 14 – February 18

QUESTION:
- *What Does the PPI (Producer Price Index) Tell Us?*

INDICATORS:
- GD, SF, CC, BY

SITUATION:

The core Producer Price Index (PPI) for January came in much higher than the expected 0.2% at 0.8% on Friday, suggesting inflation and pulling the market lower. The Producer Price Index measures the price of goods at the wholesale level. The PPI includes the cost of crude oil as well as goods such as raw cotton, construction sand and gravel, scrap iron and steel. It includes semi-finished goods such as flour, cotton yarn, and lumber, as well as nondurable goods purchased by businesses as inputs, such as paper boxes and fertilizer. The PPI also includes finished goods such as cars and trucks if they are used in production. The PPI is widely taken to be an early indicator of inflation (as wholesalers eventually pass increased costs to retailers).

Although January's higher PPI may prove to be an anomaly, the broad market responded to higher wholesale prices with a renewed concern about interest-rate hikes. Softening that concern was the memory of Alan Greenspan's remarks in a report to Congress earlier in the week. The Fed chairman said that inflation-adjusted rates were relatively low – an utterance investors took to mean that rate hikes would be gradual. The dollar sold off on Greenspan's remarks. Dollar buyers wanted more aggressive language from the Fed because higher rates help to raise demand for dollars.

One important component of PPI, oil, moved higher this week to close Friday just under $50 at $48.50 a barrel, sending energy ETFs to all-time highs for the third week in a row. In the seven trading weeks since the beginning of the year, energy ETFs have climbed over 20%. This week, they tacked on 4%. The chart below shows the weekly returns of three energy ETFs: IYE, XLE, and IGE.

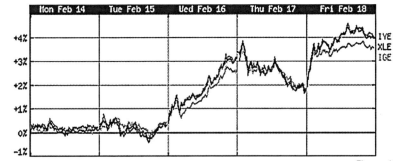

Figure 1

Some of the worst-performing sectors this week were in the financials, as investors worried that rising wholesale prices would lead to lower profits for financial institutions. The chart below shows the weekly return of three financial ETFs: IYG, IYF, and XLF.

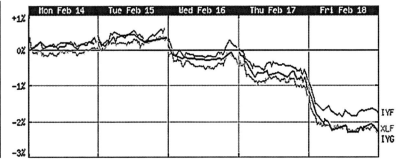

Figure 2

Fixed-income was lower, with longer-dated bonds underperforming mid-term and short-term Treasuries. The chart below

~143~

shows the weekly return of three fixed-income ETFs: TLT, IEF, and SHY.

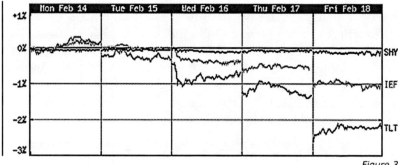

Figure 3

As the chart above shows, bonds opened lower on the PPI number Friday, with longer-dated bonds, like those in TLT, selling off most dramatically on the news. The falling dollar, which once again fell below the psychologically important $1.30 level against the euro, is also not helping fixed-income funds.

DISCUSSION:

The high PPI suggests that inflation may be a threat. The ever-buoyant oil price would seem to confirm this. Although the Fed has raised rates six times since the middle of last year, interest rates have not yet risen fast enough to prevent wholesale prices from rising. Unless the Fed begins to raise rates more aggressively, or oil falls back, inflation may happen, even accompanying higher interest rates. If, despite higher rates, inflation becomes a reality here, there are a number of possible moves that might help insulate a portfolio from inflation-related wealth erosion. The best way to hedge a portfolio against inflation is through diversification – a mix of equity, fixed income, precious metal and real estate. If inflation becomes an issue, I would expect these sectors to become popular with investors and provide opportunities.

But before we worry about inflation, reflect that as recently as 2003, many U.S investors were worried about *deflation*. Investors feared that the U.S. economy, characterized by falling prices for technology products, sky-high productivity growth and spare capacity, would begin to resemble the moribund economy of the Japanese. In Japan, it was thought, interest rates, already at 0.025%, meant that economic stimulus had failed.

Back in 2003. the problem with a deflationary climate was that businesses allegedly lacked pricing incentives, and an already sluggish consumer was further encouraged by falling prices to postpone spending. But hardly two years later, with U.S. productivity growth slowing, oil hitting all-time highs, and the dollar collapsing, worries about deflation in the U.S. economy are history.

The chart below shows PPI over a 5-year period:

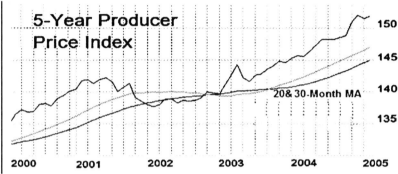

Figure 4

The jagged black line shows wholesale prices on a 5-year basis. The two smoother lines are the 20-day and 30-day moving averages. On the bottom of the chart is the calendar year. On the right scale of the chart are wholesale price levels (the year 1982 is the baseline, set at 100).

As Figure 2 shows, in early 2000, wholesale prices had drifted 35% higher than they were 18 years earlier in the base year, 1982. In the five years since 2000, wholesale prices have risen 12%. But wholesale prices actually declined sharply in 2001 and remained low in 2002. This drop caused a temporary speculation that the United States was on the verge of a Japan-style price deflationary recession. This situation lasted about two years – to the beginning of 2003. Since 2003, producer prices have risen sharply, almost 7% in two years, and the economy improved and there is no longer talk of deflation.

In the fixed-income world, a good inflation hedge is a position in inflation-adjusted Treasury bonds like those held in the ETF TIP portfolio. TIP holds bonds with an average maturity of 11 years and is indexed to the Consumer Price Index (CPI), which historically has risen faster than the PPI. Over the past year, TIP bonds have improved, but not significantly more than other non-inflation-indexed bonds. The chart below shows the 1-year performance of TIP.

Figure 5

Other ETFs that might do well if inflation picks up include those specialized in paying dividends. Dividends become attractive in periods of inflation because they help offset higher interest rates. Currently, there are two ETFs focused on dividend-paying stocks. The first is the iShares Dow Jones Select Dividend Index (DVY). The second is the Powershares High Yield Dividend Achievers (PEY). The chart below compares the 1-year returns of these two ETFs.

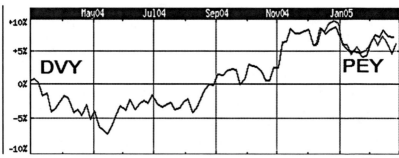

Figure 6

Another option, of course, is gold. If inflation really becomes a problem, then Gold Shares (GLD) will likely outperform. Shares are each worth about 1/10th of an ounce of gold, and track the price of gold. The chart below shows the 3-month price chart of GLD.

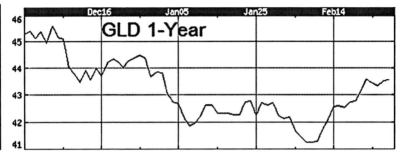

Figure 7

A prime cause of inflation in the United States is a weaker dollar because this raises the cost of imported goods (and often domestic substitutes for those goods as well). If foreign economies slow, or if the dollar strengthens for other reasons, this will help hold inflation down. On the other hand, if the dollar continues to slide, this will add to inflation fears. Owning broad-based foreign ETFs denominated in dollars but holding firms valued in other currencies will therefore provide a partial hedge against domestic inflation.

U.S. inflation could drive interest rates and the dollar higher. Investors in foreign stock are exposed to more than just currency risk, they are also exposed to the underlying foreign economy. Nevertheless, diversifying out of the U.S. markets can be an important part of a strategy aimed at safeguarding capital from domestic inflation. The chart below shows the 1-year performance of one of the foreign diversified funds, EFA.

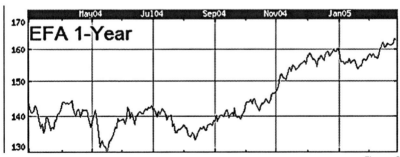

Figure 8

When it comes to battling inflation, none of these solutions is great. TIP is potentially vulnerable to rising interest rates. DVY pays a dividend but introduces broad market exposure. As seen in the first week of this year, GLD is vulnerable to a rising dollar. EFA carries foreign exchange risk.

Nonetheless, if inflation picks up, these funds can be expected to appreciate. Concern about inflation is a good reason for keeping the tech short. For now, no change.

PORTFOLIO:

ETF	LW	TW	P/L
XLE	12	12	↑4.0%
IDU	6	6	↓0.4%
QQQQ	-1	-1	↑0.9%
IGM	-3	-3	↑1.0%
EWZ	1	1	↑2.7%
Portfolio +1.7%			

LW: Last Week | TW: This Week | P/L: Profit/Loss

Week 35: February 22 – February 25

QUESTION:
■ *Fear or Greed?*

INDICATORS:
■ EN, BY, IP, FA, TA

SITUATION:
The broad market overcame a lower dollar and higher crude oil prices to end the week with gains. As the chart below shows, the recovery started late Thursday.

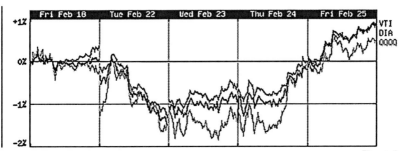

Figure 1

The chart above shows how the Vanguard Total Market Index (VTI), DIA, and QQQQ all traded in parallel, falling midweek but ending slightly positive.

Oil futures traded to $51.49 a barrel Friday, a 4-month high, on speculation that OPEC would seek to maintain higher prices. Energy ETFs responded by ending yet another week with phenomenal returns. The chart below compares the weekly returns of XLE, IYE, and the more globally focused IXC.

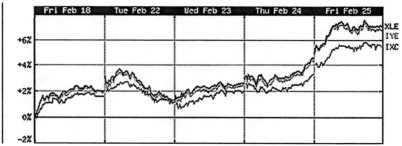

Figure 2

After six weeks of consecutive highs, Brazil made yet another new high this week, despite warnings from its central bank that rates would be moving higher. EWZ is up almost 20% in February (14% in 2005). One fundamental reason is that the strength in the Brazilian economy is due to rising iron ore prices. Another thing working for the dollar-denominated EWZ is Brazil's currency, the real, which, unlike major currencies, is gaining against the dollar. Mexico is also up, benefiting from high oil prices and strong fourth-quarter earnings. The chart below shows this week's returns of Brazil's EWZ, Mexico's EWW, and Latin American ILF.

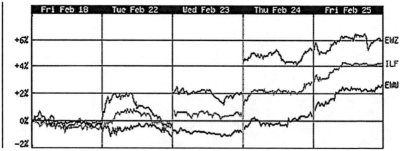

Figure 3

DISCUSSION:

Fear or greed? Part of the market is at limit points. Energy prices, of course, are at all-time highs. The dollar is collapsing again to multi-year lows. Brazil and Mexico are at all-time highs.

The broad market is also nearing new highs. The chart below shows the 1-year return of the DIA and SPY.

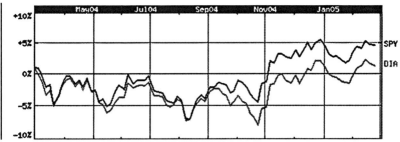

Figure 4

In the last 52 weeks, the highest close for the DIA was on December 28 of last year, when it closed at $108.40. Friday's close was 108.36, just 4 cents less. Before December 28, the previous high close on the DIA was way back at the end of 2001. Friday's close on the SPY was 121.43, a new 52-week high. Previous highs on the SPY were back in early 2002. This technical situation is important, because traders watch to see if the broad market moves to new highs. If it does so convincingly, then such a move is often followed by new buying. On the other hand, if these ETFs approach new highs but do not move convincingly above them, for many traders this will suggest an opportunity to sell.

The last of the three main index ETFs is the QQQQ.

Figure 5

As the chart above shows, unlike DIA and SPY, the QQQQ is not sitting close to its highs. If it were, the technical situation would look a lot better for the bulls, because that would suggest that the whole

~151~

market was strong. The lagging QQQQ suggests that the market will not have strength to go higher unless something definitive happens, such as a strong move above the current range. So, even though QQQQ is up on the week, it looks as if the bears retain the edge, for now.

▲ *A move near new highs is often accompanied by increased volume and increased volatility. Any significant news item, positive or negative, can be especially influential at these levels, at least for the short-term direction of the market.*

Of course, given the number of things to worry about and the market's recent indecisiveness, this difficult, nerve-racking situation might also persist for a longer period.

IGW, meanwhile, finally hit my target Friday, closing at 54.41. Although technology looks weak, IGW is improving, and this is the pre-described technical level for entry.[1]

The energy sector has been up for the last seven weeks, and has made new all-time highs every week in February. These gains accelerated this week, tacking on another 5-7% in the last five days alone. This week's strong gains make energy easily the best-performing sector in 2005, with gains of around 25% for January and February alone.

▲ *Often buyers get ahead of themselves, and the last stages of a climb are always the steepest. The recent increases in the size of the returns in the oil sector send a warning flag that a near-term top may be forming.*

The weekly returns for XLE over the last four weeks are: up 5.2%, up 3.5%, up 4%, and this week up 5%. This does not mean that these ETFs will not continue to go up. They may. And they may go up even more steeply next week. They may also continue to outperform the broad market. But there will also be a pullback at some point, and this can feel brutal, especially with a big position. It is probably OK to lighten up a little here. Certainly, there is no way that energy sector can continue with this kind of return for the rest of the year. So on a percentage basis, the biggest opportunity has already happened.

The next move in oil is a question of fear and greed. A smart trader here thinks about what is more likely to win out. I'm going to

---

1   See Chapter 30.

tilt slightly to fear, and lighten up just a little on the oil ETFs. Next week, I'll lighten up a little more. I don't want to lose this position. There is none like it in this market. I want to add the SPY on these all-time highs. I am opening a position in the SPY.

PORTFOLIO:

ETF	LW	TW	P/L
XLE	12	10	↑5.0%
IDU	6	6	↑0.6%
QQQQ	-1	-1	↓0.7%
IGM	-3	-3	↓0.3%
EWZ	1	1	↑6.3%
IGW	0	1	
SPY	0	2	
Portfolio +2.4%			

LW: Last Week | TW: This Week | P/L: Profit/Loss

Week 36: February 28 - March 4

QUESTION:
- *Biogen in Trouble: Sell the Sector?*

INDICATORS:
- GD, FA

SITUATION:
Stocks and bonds ... ETFs, almost everything went up this week. The big news for most investors came Friday with the February employment report, which by suggesting both increased hiring and higher unemployment, pleased nearly everyone. Equity buyers looking for a strong economy liked the nonfarm payrolls number, at 262,000, significantly stronger than the forecast of 225,000. This made February's job growth the best since October and helped push a broad market struggling for direction to new 3½-year highs. But the overall unemployment rate moved higher than expected, to 5.4%. Equity investors saw no evil here, largely interpreting this to mean that unemployed workers are returning to job hunting. But this higher number helped reassure bond investors, who reasoned out that a higher overall unemployment figure might slow rate hikes. Bonds had sold off in advance of the employment report, out of worry that a good report might motivate the sluggish Fed to raise rates more aggressively. Bonds rallied. The first chart below shows how the broad market equity ETFs DIA and SPY traded. The

second shows how the long-term bond fund TLT and the mid-term bond IEF traded.

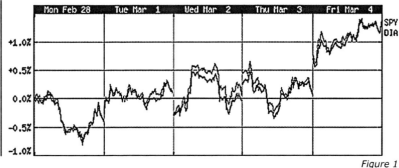

Figure 1

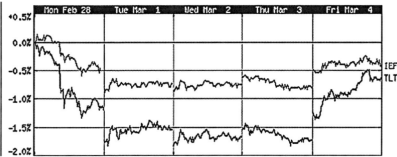

Figure 2

In Figure 2 above, the sharp move Friday shows the results of the employment report. The drop in TLT and IEF from Monday night to Tuesday morning reflects a dividend payout.

One of the few dark spots clouding the sky this week was Biogen (BIIB), a major biotech firm. Biogen pulled the plug on its multiple sclerosis drug Tysabri and fell over 40% on the news, wiping out close to $8 billion in market capitalization. The biotech sector ETF IBB dropped 4% in one day, as the whole sector fell in sympathy. Before Monday, Biogen was on an uptrend. After Tysabri's failure to obtain approval, Biogen fell dramatically, more than wiping out its gains for the year. The 52-week chart below tells this story: On Friday of last week, Biogen traded above $65. On Monday, Biogen collapsed to below $40 a share.

Figure 3

DISCUSSION:

William Haseltine, the founder of Human Genome Sciences (HGSI) and one of the most visible spokesmen for the promise of biotechnology, has turned negative on the biotech industry.[1] Haseltine used to talk about a new era in medicine in which tailor-made gene-based products would replace drugs, and "regenerative medicine" based on biotech research would result in a dramatically increased human lifespan. But in an interview with Fortune magazine Haseltine said, "I see great inefficiencies in pharma and biotech firms. We have all these opportunities but insufficient structures to pursue them." Because biotech was supposed to save big pharmaceutical companies by providing promising new compounds for development, Haseltine's switch, from selling the biotech century to grouping biotech companies with the bureaucracy of big pharma, is another arrow in the side of the biotech investor.

Things certainly looked different when the Merrill Lynch Biotech HOLDR's Trust (BBH) was introduced in late 1999, when the ETF

---

[1] In general when investing, I have stayed away from authorities, because the market eludes and punishes so-called expertise. In fact, I've found that "unless the source of the statement has extremely high qualifications, the statement will be more revealing of the author than the information intended by him." This is a quote from Nassim Taleb's *Fooled by Randomness*, pp. 223-224. The author calls this phenomenon "Wittgenstein's Ruler." Although Haseltine quit Human Genome Sciences, in this case, I think that his authority is sufficient to communicate relevant biotech sector experience. It is unclear, of course, to what extent the market already reflects this view.

was prized for the diversity it provided to investors. It soared almost 150% in its first six months. Then the tech bubble burst, and in the next two years, BBH fell 70%. Since then, as the chart below shows, BBH has recovered somewhat.

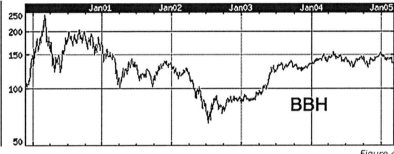

Figure 4

BBH performed well in 2003, nearly doubling. But last year, BBH and the newer biotechnology fund IBB went nowhere, landing among the worst-performing ETFs in 2004. This year, BBH and IBB are drifting dramatically lower. Part of the explanation is that people like Haseltine are no longer bullish on biotech. Also, biotechnology companies are no longer market leaders in the way they were during the run-up of the Nasdaq, when they sold at a premium for providing leadership. Today, the leaders are in energy, and money continues to rotate out of biotech. Over the last year, this process has accelerated. The 52-week chart below shows the mirror-like performance of IBB, BBH, and an energy ETF, XLE.

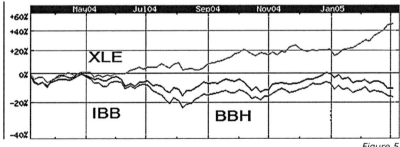

Figure 5

As the chart shows, over the last year, oil and biotech have consistently moved in opposite directions. This indicates that sector rotation is part of the story of biotechnology's underperformance, as is the relatively high beta of biotech ETFs. This most likely means that when the broad market falls on higher oil, biotech falls even farther.

On a more fundamental level, plaguing the biotechnology sector is the abundant supply of biotech names. Because of the urgent need to raise cash for research and the long development cycles in biotech, there has consistently been a new supply of biotechnology IPOs. Biotech accounted for the largest percentage of any sector going public last year. Although IBB and BBH do not invest materially in these new companies, new deals compete with more established companies for investor dollars. While the market pumps out new IPOs, IBB and BBH are falling almost daily and now flirting with new 6-month lows.

IBB may be a buy here. Biogen's Tysabri may come back on the market, and even if it does not, other important products will provide new growth and profits for biotech firms.[2] Biotechnology is an emotional boom-and-bust sector – hot in the late 1980s, cool in the early 1990s, hot in the late '90s, and now cooling again. The Biogen IDEC news last week might provide a good entry point to buy the bust.[3] But given its current direction – down – a long-side position in IBB is very counter-trend and clearly the kind of trade I am not prepared to make.

PORTFOLIO:

Like biotech, technology failed to participate in the overall optimism. Given this broad market strength, failure to participate is

---

2   Avastin, the new drug from Genentech (DNA), Truvada from Gilead Pharmaceuticals (GILD) and Erbitux from ImClone (IMCL) are the probably the most important. But also in the pipeline are Myozyme for Pompe's Disease from Genzyme (GENZ), Revlimid for Myelodysplastic Syndrome (MDS) from Celgene (CELG) and ABF-EGF for colorectal cancer from Amgen (AMGN) and Abgenix (ABGX), and Telcyta for ovarian cancer from Telik (TELK). Medarex (MEDX) has monoclonal antibody based products in various stages with Pfizer (PFE) and Bristol Meyers (BMY).

3   Of the two biotechnology ETFs, IBB is more diversified and has greater exposure to smaller companies. Because it is designed to track the Nasdaq Biotechnology Index, IBB does not include Genentech, which trades on the NYSE. Most of the difference in the performance of these ETFs is traceable to the performance of Genentech.

notable. The QQQQ continues to lag behind the DIA and the SPY, and to look for direction. But sectors within technology are not performing alike. The chart below compares networking (IGN), semiconductors (IGW), and software (IGV).

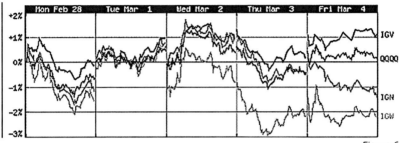

Figure 6

As the chart shows, these ETFs are all over the map. Weakness in technology when the rest of the market is strong underscores the sector's weakness. But technology direction is not unambiguous. I want to see all tech ETFs moving lower in tandem before increasing the tech short here. Biotech is different. Add to the biotech short on the back of the BIIB news. And continue to lighten up on oil.

ETF	LW	TW	P/L
XLE	10	9	↑1.7%
IDU	6	6	↑1.7%
QQQQ	-1	-1	↑0.3%
IGM	-3	-3	↓0.3%
EWZ	1	1	↑0.6%
IGW	1	1	↓2.3%
SPY	2	2	↑1.3%
IBB	0	-1	
Portfolio +1.1%			

LW: Last Week | TW: This Week | P/L: Profit/Loss

Week 37: March 7 – March 11

QUESTION:
■ *How Important Is It to Be Right?*

INDICATORS:
■ GD, EN, BY, CC, IP

SITUATION:
January's $58.3 billion trade deficit, the second-highest monthly deficit in U.S. history, capped off a week of declines for the broad market. The deficit reflected higher January oil prices as well as a surge in clothing imports from China. The markets were rocked this by near record-high oil, a weak dollar, inflation worries, and rate hikes. The chart below compares the SPY with gold (GLD).

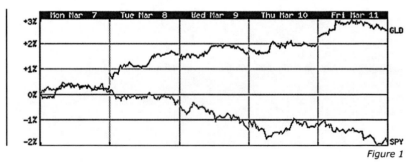

Figure 1

The chart above shows that the return on the equity benchmark

SPY was the inverse of gold, one of the most conservative investments.

The sellers who drove fixed-income down last week came back in force and sold bonds. The long bond TLT, holding Treasury bonds with maturities of over 20 years, dropped 2.5%. IEF, a Treasury bond portfolio holding intermediate-term Treasury debt with maturities of 7 to 10 years, dropped 1.5%.

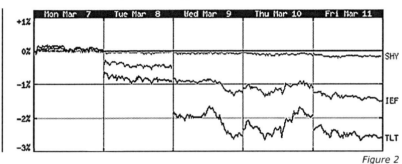

Figure 2

Brazil (EWZ) has been a star performer and market darling. But things reversed this week. The lower U.S. bond prices in Figure 2 above hurt Brazilian stocks. The promise of higher U.S. Treasury yields means that emerging market areas such as Brazil could be starved off, as lenders are drawn north. The chart below compares EWZ with Mexico (EWW), and the pan-Latin American fund (ILF).

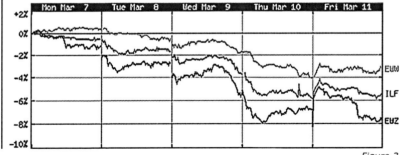

Figure 3

The similarity in the shape of the curve of TLT in Figure 2 and EWZ in Figure 3 above shows the linkage between the Brazilian equity market and U.S. bond market.

Most broad market large-cap ETFs slightly outperformed mid-cap funds. Small-caps significantly underperformed. The chart below shows the large-cap DIA, the mid-cap MDY, and the small-cap IWM.

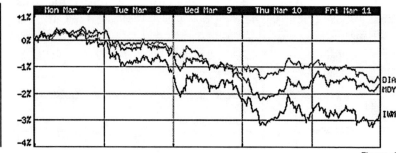

Figure 4

DISCUSSION:

The portfolio took a massive hit this week. I have too many positions, and everything went wrong. When this happens, I downsize out of the losers and strengthen the winners (if there are any) without procrastinating or rationalizing the loss. This is something a trader has to do over and over again, and in a sense this is his job – to get out when things go wrong. This means not becoming attached to any position. This is why trading is, in a sense, inhuman. Love and loyalty, so important in life, can be disastrous for the trader. To quote Jesse Livermore, "A man does not swear eternal allegiance to either the bull or the bear side. His concern must be with being right."[1]

I've become too human, too complacent about the oil position – I didn't sell enough. Steadily rising oil prices made energy the hottest place to be in January and February. But the sector is increasingly volatile. This week, the key energy fund XLE lost 7% between noon Wednesday and early morning Thursday. The chart below compares the performance of three energy ETFs, XLE, IGE, and the more globally focused IXC.

---

[1] Lefèbre, Edwin, *Reminiscences of a Stock Operator*, p. 181. Sadly, Livermore, one of the early proponents of technical analysis, was not capable of following his own advice. One of the greatest traders of all time, Livermore shot himself in the Sherry Netherlands Hotel in 1940.

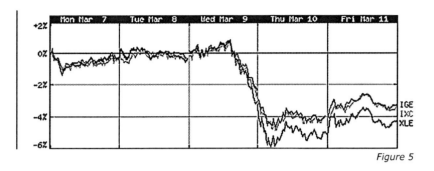

Figure 5

PORTFOLIO:

The fun is over for Brazil and the EWZ trade. There is too much volatility here to have a position in this fund. I'm cutting back the utility position. And if oil is taking off and the whole tech sector is collapsing, IGW can't survive as the lone maverick. I'm getting smoked. I'd like to hold on until I get even, but this is the typical folly. I need to try to trade more automatically, buy this back here and take my loss before things get worse. I'll close IGW, and wait for it to cross 54 again before buying back.

The only things really working here are the short tech and biotech and positions. Increase these. Everything else gets cut.

ETF	LW	TW	P/L
XLE	9	6	↓4.7%
IDU	6	4	↓1.5%
QQQQ	-1	-1	↑0.9%
IGM	-3	-3	↑1.2%
EWZ	1	Closed	↓6.6%
IGW	1	Closed	↓1.4%
SPY	2	1	↓1.9%
IBB	-1	-2	↑2.7%
Portfolio -2.0%			

LW: Last Week | TW: This Week | P/L: Profit/Loss

Week 38: March 14 – March 18

QUESTION:
■ *What Is the Easy Trade?*

INDICATORS:
■ EN

SITUATION:
The broad market fell this week, as retail sales numbers disappointed, and oil futures made record highs. Oil touched $57.50 a barrel Thursday, before falling back to trade at $56.72 late Friday. The chart below compares the 1-week performance of the DIA, SPY, and QQQQ.

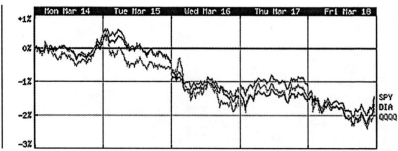

Figure 1

With supply static, oil is moving higher, largely on increased demand. This was apparent again Thursday, as the price spike in the

~164~

oil contract came just after OPEC's discussion to raise its quota to a record high of 27.5 million barrels per day. With OPEC pumping more than ever and Saudi Arabia the only cartel state effectively capable of delivering a production increase, analysts have targets of $60, $65, and even $80 a barrel. Energy ETFs followed all-time highs in crude to new all-time highs of their own. The chart below compares the weekly returns of XLE and IGE.

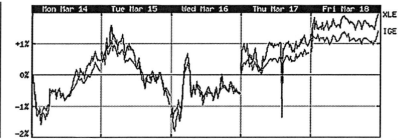

Figure 2

Internationally, Latin America continues to struggle. This week, it was Mexico (EWW) that sold. Brazil (EWZ) was last week's worst performer, down about 6%. On Wednesday and Thursday Mexico and Brazil went in sharply opposite directions, to close the week more than 4% apart. The chart below shows this week's returns of EWW and EWZ.

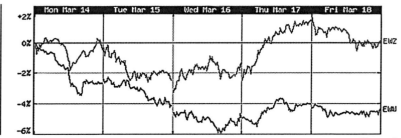

Figure 3

⚠ *There may be an opportunity to trade EWZ against EWW here on a mean reversion basis. According to a mean reversion strategy, the prices of EWW and EWZ should revert to a mean return. So a trader would arbitrage EWW and EWZ, waiting for a week with a significant diversion, and then buy the lower index and sell short the higher. This strategy would have worked well this week, as last week, Brazil* underperformed Mexico, and this week, it was the reverse: Mexico underperforming Brazil.

DISCUSSION:

There may be a way to trade Mexico against Brazil, but what is the easy trade? What has been consistent since the beginning of the year? Higher oil and lower technology. The two charts below show the returns of the cubes as compared with software (IGV), networking (IGN), and semiconductors (IGW). The upper chart compares these four tech ETFs on a weekly basis. The lower chart compares them on a 6-month basis.

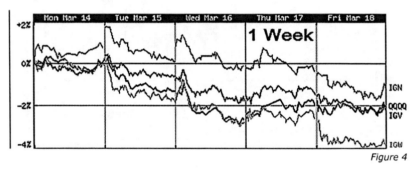

Figure 4

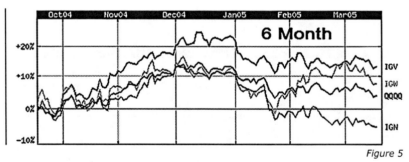

Figure 5

Technology is collapsing. QQQQ is now trading at 4-month lows, and IGN at 6-month lows. The whole sector ended the week on its lows. Next week's PPI and CPI might help tech if they come in lower than expected. But given the high oil prices, PPI and CPI, if anything, look vulnerable. I want to add more to the short tech position.

It is time also to get out of the rest of the SPY trade. I bought that just after the S&P 500 hit a 3-year high. Since then, SPY has not traded higher. The SPY is down two weeks out of three since then. This position isn't working, and the market looks poised to go lower.

PORTFOLIO:

ETF	LW	TW	P/L
XLE	6	9	↑2.6%
IDU	4	4	↑0.4%
QQQQ	-1	-2	↑1.8%
IGM	-3	-4	↑2.1%
SPY	1	Closed	↓1.2%
IBB	-2	-2	↑0.3%
Portfolio +1.0			

LW: Last Week | TW: This Week | P/L: Profit/Loss

Week 39: March 21 – March 24

QUESTION:
- *What Does U.S. Inflation Tell Us About Foreign ETFs?*

INDICATORS:
- BY, CC

SITUATION:
The big move of the week came late Tuesday, following the Federal Reserve's routine 25 basis-point rate hike. Greenspan did more than raise rates. He warned that "pressures on inflation have picked up in recent months, and pricing power is more evident." The chart below shows this week's DIA, SPY, and QQQQ.

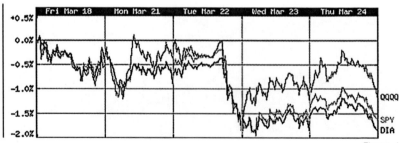

Figure 1

The sharp drop Tuesday afternoon follows Greenspan's remarks on inflation. As investors thought over the Fed's signal on inflation, bonds sold off. Tuesday's rate hike sent the yield on the benchmark

10-year note to 4.627%, its highest point since July of last year. The chart below compares the return of the long bond ETF TLT with the intermediate-term bond fund IEF and the short-term fund SHY.

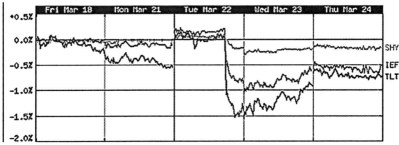

Figure 2

Many investors interpreted Greenspan's remarks to mean not only that rates would continue to go up, but that the Fed might also reevaluate more formally the risk of inflation to the economy, a move that could eventually mean more aggressive hikes. Predictably, fixed-income portfolios of longer-dated Treasuries proved especially vulnerable to these remarks.

DISCUSSION:

Although domestic ETFs were hurt by Tuesday's sell-off, the damage was minor compared to some international issues, particularly in countries where there is intense competition for investor dollars. The Australia Index (EWA), with its exposure to the Australian dollar, crashed as did Brazil (EWZ). The chart below compares the 5-day return of EWZ and EWA.

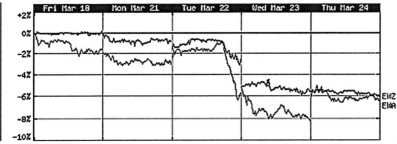

Figure 3

With its relatively high interest rates, Australia has had a yield advantage over U.S. government securities. The higher Fed funds rate announced Tuesday eroded that advantage. The Australian dollar and the Australian market moved sharply lower. Not shown in the above chart, but also affected by the threat of U.S. inflation is the iShares MSCI South Africa Index, also down about 6% on the week.

PORTFOLIO:

The relative strength of the QQQQ is concerning. In past weeks, the QQQQ has fallen more sharply than the more industrial broad market DIA and SPY. This week, it was the reverse. It may be time to cover the QQQQ position, and even to buy QQQQ relative strength, selling the DIA or SPY. But I do not want to change from a directional to an arbitrage position without further research.

Cut back on oil and utilities on this weakness. I would cut deeper, but this week's sell-off looks like the unusual negative event in an otherwise upward trend.[1]

ETF	LW	TW	P/L
XLE	9	6	↓4.0%
IDU	4	3	↓1.4%
QQQQ	-2	-2	↑0.7%
IGM	-4	-4	↓0.3%
IBB	-2	-2	↓1.2%
Portfolio -1.9%			

LW: Last Week | TW: This Week | P/L: Profit/Loss

---

1 See chapter 23.

Week 40: March 28 – April 1

QUESTION:
- *What Is a Fakeout?*

INDICATORS:
- EN, BY

SITUATION:
This was a volatile and difficult week for the market. The market moved on oil, which traded Friday to a new all-time high of $57.27 a barrel. This took the broad market to 3-month lows. The chart below shows the performance of the QQQQ, DIA, and SPY.

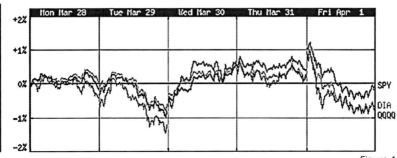

Figure 1

Oil ETFs sold off mid-week but moved up on a spike in gasoline futures and a bullish report out of Goldman Sachs predicting that crude would go to $105 per barrel. Energy ETFs were up about 4%

and close to their all-time highs, set at the end of last month.

The key moment for ETFs with energy exposure came midday Wednesday, when oil futures dropped to $52.50 after the U.S. Energy Information Administration reported a 1-week increase of 4.1 million barrels, pushing the overall inventory to 309.3 million barrels, its highest level in almost three years. With oil at record highs, this is all it took to spark a sell-off. But the oil bulls took advantage of the news to buy. The chart below shows Wednesday's inflection point in XLE.

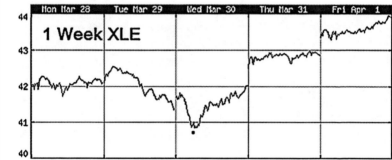

Figure 2

At the inflection point indicated early Wednesday in the chart above, over 55,000 put options traded on XLE. This brisk trading was likely the result of bullish investors selling out-of-the-money put options to help finance their purchase of the underlying fund.

DISCUSSION:

A fakeout is when a trader finds an explanation for market movement and enters into a position, only to find that the movement does not develop and the asset reverses direction. Three fake-outs. The first is the phony sell-off Wednesday in oil service ETFs, shown in Figure 2 above.

The second concerns ETFs that sold last week as investors worried that the higher Fed funds rate would erode the yield advantage that some foreign markets had over U.S. government securities and take investment dollars out of these markets, bringing them to the United States. The chart below shows last week's performance of EWZ and EWA.

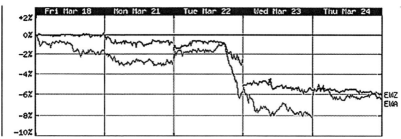

Figure 3

While internationally most ETFs were little changed on the week, big movers from last week, Brazil (EWZ) and South Africa (EZA), reversed. Australia (EWA) traded sideways. The chart below compares the weekly returns of EWZ and EZA with a global benchmark ETF, the iShares Global S&P 100 (IOO).

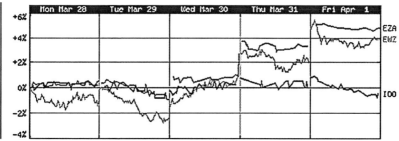

Figure 4

On the fixed-income front, rising rates usually mean a bear market for bonds. The chart below shows the performance of TLT, IEF, and SHY following the rate hike on Tuesday of last week.

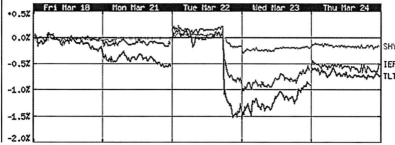

Figure 5

Last Tuesday's chart looks like a cliff, with bonds flat before the

Fed announcement and getting crushed almost immediately. Except for this sharp 15-minute correction, bonds were resilient. The uptrend that began last Wednesday carried into this week, helped by the disappointing Q4 GDP. Investors fled to bonds. The chart below compares the weekly return of TLT and SHY with the aggregate fund AGG (holding Treasuries and corporate bonds).

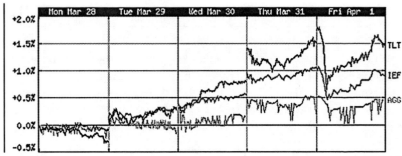

Figure 6

As the above chart shows, the rally in fixed-income that started on Wednesday really picked up on Thursday, as the jobless report came in higher than expected. Friday's cliff-like sell-off did not keep bonds from their first weekly gain since the beginning of February.[1]

PORTFOLIO:

Oil ETFs continue to remain easy to own, and with oil at new highs, adding to the oil position is self-evident. Add XLE.

The position in QQQQ didn't fall as much as I expected given this oil spike. Still, I don't expect the QQQQ to spike up in one week, so for now, I'll just hang on to the short.

---

1 If the Fed continues on its apparent course, tightening at every FOMC meeting, these funds do face further downside risk in the weeks and months ahead. But because of their fat coupons – the high yields these bond funds provide – as well as the continued domestic and foreign demand for bonds, these ETFs can be notoriously difficult and expensive to short. Also important is that fixed-income ETFs continue to do remarkably well. In the end, bonds remain both difficult to own and difficult to short.

ETF	LW	TW	P/L
XLE	6	7	↑4.5%
IDU	3	4	↑1.7%
QQQQ	-2	-2	↑0.2%
IGM	-4	-4	↑0.3%
IBB	-2	-3	↑4.1%
Portfolio +2.4%			

LW: Last Week | TW: This Week | P/L: Profit/Loss

Part IV, Q2 2005

~

Week 41 - Week 52

Week 41: April 4 – April 8

QUESTION:
- *What Happens When Oil Collapses?*

INDICATORS:
- EN, CC

SITUATION:
OPEC said it would step up oil supply and allow U.S. stockpiles to rise. After hitting record highs just last week, oil tumbled 7% in five days on the news to close $53.32 a barrel. Below is the 5-day price chart of the benchmark oil futures contract.

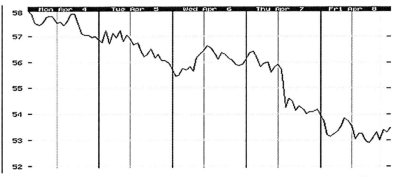

Figure 1

Oil service ETFs followed crude lower, shedding 3% on the week. The chart below shows the return of XLE, IYE, and IXC.

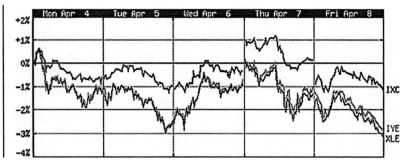

Figure 2

When crude is up, oil ETFs, utilities, and REITs have been usually higher. All other sectors are stagnant to lower. When crude is down, the market as a whole moves up. As crude fell this week, the Vanguard Healthcare (VHT), iShares Dow Jones US Healthcare (IYH) and Biotechnology (IBB) were big winners.

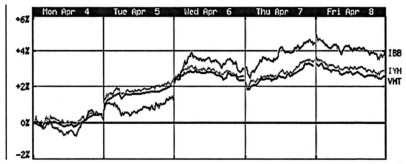

Figure 3

These ETFs moved higher despite negative news of an FDA decision to force Pfizer (PFE) to pull its embattled arthritis drug Bextra off the market, a move that has led to speculation about further regulation of health care and pharmaceuticals.

The broad market also celebrated the collapse in oil. The party lasted until Friday morning, when profit-taking hit Wall Street and investors sold the market going into the weekend. The chart below compares the large-cap SPY with the mid-cap MDY and the small-cap IWM.

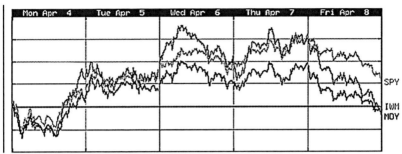

Figure 4

The chart above shows that the Friday sell-off was particularly acute for small- and mid-cap funds. They gave up a week of gains in one session.

Internationally, performance was mixed. Europe outperformed. The chart below compares the returns of the three highest-volume foreign ETFs, the EAFE Index fund (EFA), the Japan (EWJ), and the Hong Kong (EWH), with the benchmark SPY.

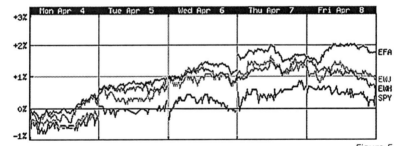

Figure 5

DISCUSSION:

Fear of higher oil prices is making the sector returns increasingly uniform. Given the bad news in pharmaceuticals, some selling might be expected. But in this case, bad news in health care and biotech was trumped by cheap oil. Biotech piggybacked on broad market strength.

Comparing Figure 1 and Figure 2, oil ETFs fell less on a percentage basis than did crude. When on Wednesday crude was flat to slightly higher, oil ETFs rallied almost 2%. While oil ETFs obviously cannot withstand unlimited selling in crude, they are holding up comparatively better than the commodity here.

Given the dramatic fall in crude, the broad market looks surprisingly weak. The chart in Figure 3 above shows the market moving up strongly midweek but unable to sustain these highs and falling severely Friday into negative territory. It looks here like the bears are spoiling optimism brought on by lower oil prices. The chart reminds me of the last day of 2004, Figure 1, in Week 27, when on Friday, Dec. 31, 2004, the bears spoiled the new highs on broad market indexes to end the week on the lows. Of course, what followed the last week of the year was a monumental sell-off in the beginning of January. We probably won't get that here. Still, looking at that chart, it may be a good idea here to go against the trend and add to the short tech positions, and to oil longs.

PORTFOLIO:

The portfolio was on the wrong foot this week. The identified trend is lower prices for technology ETFs and higher prices for oil ETFs. This week, the market did the reverse. Does this mean the portfolio here should be reversed? No. This would be calling a change in a long-term trend and trying to make market forecasts. The way this is set up, the portfolio is doubly hurt when oil falls, because it loses money both on the oil service ETFs and on the tech rally. Again, when this happens, I reduce my overall exposure to the market and add to what works. What worked this week? Utilities. Add to the utilities position. Cut back on oil longs, technology shorts. Ending the week on its high. The biotech short looks dangerous. Time to cover.

ETF	LW	TW	P/L
XLE	7	5	↓2.8%
IDU	4	4	↑0.7%
QQQQ	-2	-1	↓1.2%
IGM	-4	-3	↓0.3%
IBB	-3	Closed	↓3.8%
Portfolio -1.6%			

LW: Last Week | TW: This Week | P/L: Profit/Loss

Week 42: April 11 – April 15

QUESTION:
- What Does April 15th (Tax Day) Tell Us?

INDICATORS:
- EN, GD, SF, IP

SITUATION:
April is the cruelest month. Everything the market gained last week, it gave back this week – and more. The broad markets broke down, with technology leading the collapse. The QQQQ lost almost 6% in three trading sessions. Friday was a monolithically bad day. On Friday, the market fell almost 2% on heavy (1.5 times normal) volume. The DIA lost over 1.5% on Friday, its worst single day in two years. The chart below compares DIA, QQQQ, and SPY.

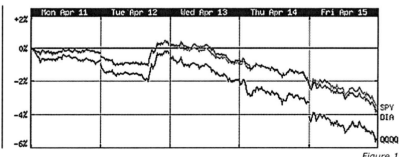

Figure 1

Here is the anatomy of the sell-off: The market started selling Tuesday, when, as if it were looking for a $59 billion trade deficit for February, high oil prices helped move February's deficit to a record $61 billion. As the above chart shows, buyers came in late Tuesday, scaring the short sellers. But with Wednesday's release of disappointing retail sales data for March, the selling resumed in earnest. Expected at 0.8%, March retail sales came in at 0.3%. Retail sales ex-autos were even weaker: The market expected 0.5%, but it got just 0.1%.

Then came bad news from several bellwethers, seeming to confirm a possible economic slowdown in the months ahead. Markets continued lower in Asia overnight Wednesday, on bad news out of Canon and other key tech companies. This was, perhaps, a premonition for the United States, because on Thursday, Apple Computer (AAPL), a recent bellwether for the tech sector, disappointed, and technology ETFs continued to sell off sharply. The chart below compares three technology sector ETFs, the software (IGV), networking (IGN), and semiconductors (IGW) with the more broad-based technology IGM.

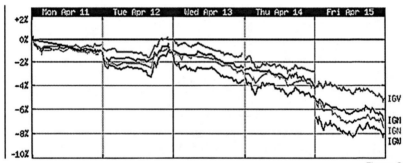

Figure 2

It was not just technology that sold. On Thursday, JP Morgan Securities warned that the all-American standard, General Motors (GM), which missed earnings last week, may now see its debt downgraded. Because GM and many of its suppliers were already crossover debt, this will potentially add $200 billion in new debt to the high-yield sector.

Friday brought news of sharply higher import and export prices. And Friday, it was International Business Machine's (IBM) turn to announce disappointing profits. IBM's lower profits seemed to confirm the suspicion that technology spending would fall in 2005, and tech stocks dropped further. At this point, the broad market was in free-fall, triggering technical selling as indexes moved to 5- and 6-month lows.

As is often the case when markets are selling off sharply, large-cap ETFs mostly outperformed small- and mid-cap funds. The chart below compares the returns of the small-cap fund iShares Russell 2000 Index (IWM) with the mid-cap iShares Russell 2000 Midcap Index (IWR), and the large-cap iShares Russell 1000 Index (IWB).

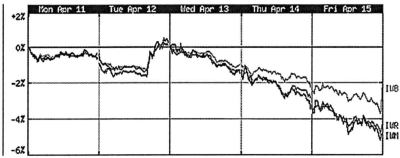

Figure 3

Hong Kong (EWH) was one of the best performers among major nondomestic ETFs, and the volatile Brazil (EWZ) among the worst. EWZ continues to trade off concerns of higher U.S. interest rates and the outlook for steel and mining industries. A 2-month high on the dollar against the euro did not help EFA, an international ETF with European holdings. The chart below compares the weekly performance of the most important international ETFs: EWH, EWZ, EFA, Japan's EWJ, and Mexico's EWW.

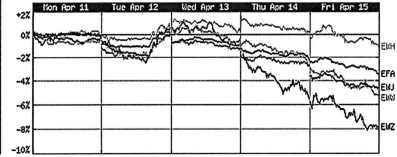

Figure 4

Winners were hard to find this week, but there were a few pockets of strength, notably in fixed-income and REITs. Health care and biotech were lower but outperformed. The chart below compares the returns of three bond funds, TLT, IEF, and SHY, with IYR, a REIT fund.

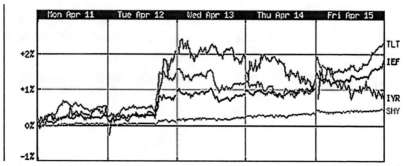

Figure 5

Comparing the charts in Figure 1 and Figure 5 above, it becomes clear that during the period of the most serious selling, from the opening on Wednesday, April 13, to the close on Friday, April 15, bonds improved, but REITs did not improve.

DISCUSSION:

Normally, I would start cutting back on things that are not working and adding to things that are working, but this week's sell-off is so dramatic – with a 6% drop in the Nasdaq, 4% drop in the Dow and S&P, that this is not just a question of my portfolio or positioning, but a question of the broad market. This kind of panic usually doesn't happen more than a few times a year. When the market elephants go this far this fast, I think we can expect an equal and opposite return in the next couple of weeks. After IGW falls 8% in one week, this is definitely time to take profits on any remaining short positions in technology – in this case, unfortunately, only the small tech short positions.

I suspect that there is less risk on the upside than on the downside, but the question becomes what sectors to get long here – technology or oil, or what? With the tech sector as weak as it is and ending the week on its lows, it is difficult to step in and buy. My favorite to buy for the bounce would be IGW because of the severity of its drop. 8% in one week – on what news? General market malaise? This is the kind of dramatic move that would be impressive and encourage a contrarian view in a single stock. In the case of a fund – which is less subject to single-stock risk – this kind of move is even more significant.

Looking back and tallying up the losses, this is the single largest weekly move in IGW since July of last year. This is also the case for QQQQ. Following that down week last year was a flat week and then a huge bounce that took the cubes and IGW right back where they were before the drop. I'm betting on that happening again.

Other than betting and speculation on investor psychology and the tendency of all markets to mean reversion, is there any other reason to be a contrarian here? Yes. One further reason to buy here is that although, for many investors, there seemed little relief from this torrent of bad news, news may be less important than seasonal factors.

⚠ *Markets are traditionally weak in the days leading up to April 15, when taxes are due. But although the week leading up to April 15 as well as tax day is historically difficult, things tend to bottom here and turn around. After April 15, markets historically rebound.*

In other words, this looks like a fundamental problem with the markets, but it may be merely seasonal. With this trend in mind, I want to exit the tech short, maintaining and adding to the oil and utilities longs and adding IGW and the cubes, with the idea of catching a rebound in tech.

PORTFOLIO:

ETF	LW	TW	P/L
XLE	5	10	↓7.0%
IDU	5	8	↓2.0%
QQQQ	-1	Closed	↑5.2%
IGM	-3	Closed	↑6.4%
IGW	0	2	
Portfolio -1.4%			

LW: Last Week | TW: This Week | P/L: Profit/Loss

Week 43: April 18 – April 22

QUESTION:
- Can Broad Market Volatility Be Used to Predict Oil Prices?

INDICATORS:
- EN, IP, TA

SITUATION:
Coming on the heels of last week's rout, this was another riotous week for the market. Broad markets moved up slightly Monday and Tuesday. But the fear of a sell-off remained. And Wednesday, dramatically higher crude prices and their negative implications for economic growth sent broad markets into a nose-dive, adding to last week's losses. Then, on Thursday, the big turnaround: After more than a week of selling, the broad market looked so oversold it wanted any excuse to rally. It found one in the often misleading weekly initial jobless claims number, which came in slightly lower than expected. Although on another week this might not even be noticed, traders used the report as a buy signal, and broad markets opened higher Thursday and never looked back. Thursday was one of the greenest days of 2005, a day with no long-side losses. It was the harmonic response to the call of last Thursday and Friday's rout. But it did not last long. Crude continued higher, moving up almost 8% on the week, and on Friday the broad market collapsed again.

The chart below compares the weekly returns of the SPY, DIA, and the QQQQ.

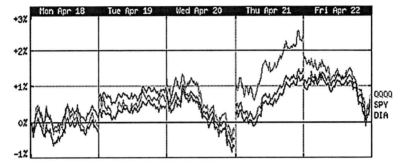

Figure 1

As the chart above shows, the relief rally Thursday did not undo last week's severe sell-off because on Friday, oil once again closed near all-time highs, at $55.39 a barrel.

Although broad market volatility and the price of crude are not often compared, with oil trading at record highs, the broad market is naturally concerned about the price of such a key input as oil. New highs on the oil contract often increase overall market uncertainty.

Known as the "fear index", the VIX measures the volatility of the ubiquitous Standard & Poor's 500. Typically, the VIX spikes in times of uncertainty or fear, such as when the market resumed trading after Sept. 11, 2001, and during the worst days of the bear market in the fall of 2002. It eases when the market is strong, or when investor confidence strengthens. The composite chart below shows the 1-year VIX and the oil contract chart on the same scale through mid-Wednesday of last week.

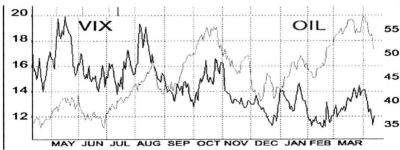

Figure 2

In the chart above, the oil contract is represented in gray.

~187~

Volatility (the VIX) is represented in black. Market volatility as a percentage is shown on the left scale, and the oil price in dollars is shown on the right-hand scale. Over the course of one year, there is no apparent correlation: Oil moved higher, volatility lower. However, on a shorter term basis – 7-day or 14-day – the oil price and market volatility track each other quite closely. Is there an opportunity here to make a bet that these two lines will converge?

DISCUSSION:

The chart in Figure 2 above indicates that, although they are tracking each other, through Wednesday of last week they moved further apart than any time in the past twelve months. This suggests that there may be an opportunity to do a convergence trade: short oil and long volatility.

If broad market ETFs continue to be worried about the price of oil, and the VIX does indeed move in tandem with oil prices, then a trader might believe that, when balanced correctly, this proposed trade could be a partially hedged position. Why? If oil does indeed move substantially higher, given that it is near a multi-year low, market volatility might be expected to move up also. If this were to happen, a trader who put on this trade would have a loss on the short oil position, but would make it up on the long volatility position. On the other hand, with volatility at multi-year lows, a further steep fall in volatility would probably not happen without a significant fall in the price of crude. In this scenario, a trader would have a loss on the long volatility position, but a gain on the short oil position.

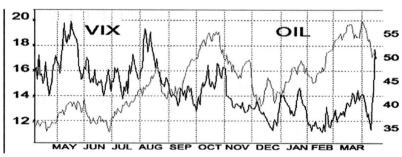

Figure 3

A look at the same 1-year chart, pulled on Monday, confirms the viability of the long volatility-short oil trade. On Wednesday, Thursday, and Friday of last week, as the market collapsed and anxiety increased, oil held steady but volatility spiked to new 6-month highs. The chart above shows the convergence of oil and the VIX.

For ETF traders, an easy way to buy and sell volatility is to trade options on the SPY. Although there is no crude ETF, XLE and IGE, though historically less volatile than oil futures, provide good exposure to oil.

PORTFOLIO:

It would have been nice to have had a little more conviction on the seasonal-post-tax-day trend. The lesson here is not to panic. Although it is usually good to exit a position on a weak chart, in very extreme markets, it usually pays to think as a contrarian. As Nathan Rothschild famously said, "Buy when the cannons are firing, sell when the trumpets are blowing." Many traders make the mistake of hearing the cannons and the trumpets too often. After this week's huge gains, it is probably time to short technology again.

Although though IGW rallied 4% and QQQQ 1%, this does not offset last week's broad losses. Time to exit IGW.

ETF	LW	TW	P/L
XLE	10	12	↑5.8%
IDU	8	12	↑2.6%
IGW	2	Closed	↑3.3%
Portfolio +3.7%			

LW: Last Week | TW: This Week | P/L: Profit/Loss

Week 44: April 25 – April 29

QUESTION:
- *What Does Earnings Season Tell Us?*

INDICATORS:
- EN, SF, GD, TA

SITUATION:
This was a difficult week for the market, and on the face of it, it looks as if the stage is set for a sell-off.

Durable goods orders, a leading indicator of manufacturing activity, skidded to -2.8% compared to the +0.3% expected. This is the kind of difference between expectations and actual results that really matters. When everybody started scrutinizing the durable goods report, a big part of the problem was a fall-off in business capital spending.[1] The bad news continued, with first-quarter GDP coming in at 3.1%, much lower than the 3.5% expected, the worst showing in two years. Even the fall of oil below the psychologically important $50 a barrel for the first time since mid-February failed to stimulate the market. The chart below shows the similar weekly performance of the DIA, SPY, and QQQQ, which all rallied on

---

1 Since 2000, in an attempt to repair balance sheets from the excesses of the 1990s, businesses are cutting spending and increasing savings. The durable goods number showed that corporations are unwilling to spend, and ever less of this massive hoard of corporate cash seems to be entering the economy.

Friday, but to levels below where they began the week.

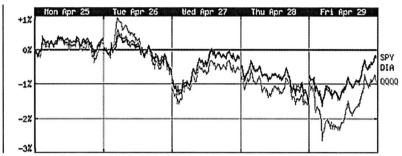

Figure 1

As investors fled equity, bonds did well. The best performer was the long bond TLT. The high-grade corporate bond fund LQD also outperformed. The TIPS Bond fund (TIP) underperformed. The chart below compares the returns of TLT, LQD, and TIP.

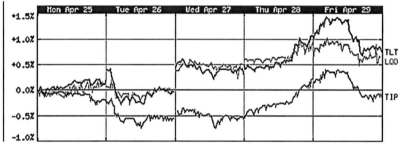

Figure 2

TIP's under-performance is notable because it has been strong all year, outperforming other fixed-income ETFs. TIP holds TIPS (Treasury Inflation-Protected Securities) bonds indexed to inflation. Unlike traditional Treasury bonds held in other fixed-income ETF portfolios, inflation causes the principal of a TIPS bond to increase. A lower TIP suggests that the market believes that the Fed is expected to be vigilant about inflation, This sentiment would nullify the advantage provided by an inflation-indexed product.

DISCUSSION:

Though things look bad this week, the market is highly volatile. Sometimes a rally starts for seasonal reasons. Below is a chart showing the performance of DIA since January 2004.

Figure 3

As the chart above shows, there are four main troughs in 2004: March, April, July, and October. Why these months? With the exception of March, these negative months are earnings season. Earnings season occurs in the month following the end of each quarter: January, April, July, October. January was a good month in 2004, but otherwise, the DIA fell during earnings season, as the market was disappointed by company results. After earnings season, the broad market perked up again. The pattern seems to have continued into 2005, with January and April both very negative months for DIA and other broad market ETFs.

As the chart above shows, although the market fall in 2004 started in early April, it continued to move lower into May, bottoming out in mid-May, and rising again until the beginning of earnings season in July, when it once again resumed a trend lower. Like the May sell-off in 2004, the July and October sell-offs lasted about six weeks. As the chart in Figure 6 shows, the sell-off during earnings season in January 2005 was briefer, lasting only three weeks.

Past results, of course, are no guarantee of future performance. But given that since March 2005 the DIA has fallen almost 10 points, or 10% (from 110 to 100), a bigger drop without a bounce than any time in the last 16 months shown in the chart, the combination of earnings season pattern makes this a good time to buy.

The threat is that the sideways market identified here is just a prelude to a negative market. Acting upon this would mean calling a change of trend. According to recent surveys, money managers are overwhelmingly bearish on the domestic economy. This, if anything, is a further good sign. Usually, when the overall sentiment is just as bearish as it can get, this is the time to buy. For many reasons, in other words, we now appear to be at the bottom of a cycle.

PORTFOLIO:

The chart below shows the stunning oil collapse, from over $55 a barrel on Monday to $49.72 by market close Friday.

Figure 4

This is precisely the buying opportunity described last week. Oil has fallen sharply. Over the last 42 weeks, whenever oil has fallen this far this fast, it has come bounding back. I have been punished for exiting my long oil sector positions. From counting, I know that oil ETFs fall one week out of three, and that they fall more than they rise. I think on this basis that oil will come back. Still, this is a big position, and I need to maintain discipline. I will sell part of the position.

In sum, decrease the oil position, which is losing, hold utilities, which are stable, and on the basis of the cyclical earnings chart in Figure 3 above, go long the DIA.

ETF	LW	TW	P/L
XLE	12	9	↓3.5%
IDU	12	12	↑0.8%
DIA	0	2	
Portfolio -0.7%			

LW: Last Week | TW: This Week | P/L: Profit/Loss

Week 45: May 2 – May 6

QUESTION:
■ *Is the Re-Introduction of the Long Bond Good for Investors?*

INDICATORS:
■ **EN, BY**

SITUATION:
The broad market bounced back from last week's losses to post gains nearly every day this week. Despite higher oil and a debt downgrade at Ford Motor (F) and General Motors (GM), fear in the market ebbed. The chart below compares the weekly performance of the SPY and QQQQ.

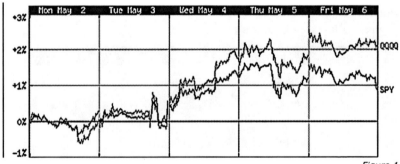

Figure 1

As the broad market advanced, the 5-day VIX, which measures

volatility on the S&P 500 Index, fell.

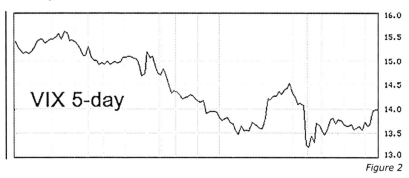

Figure 2

But the big news this week was the U.S. Treasury's announcement that it would considering reissuing the benchmark 30-year long bond, which was discontinued in 2001. The long bond was phased out because it is generally more expensive for the government than the 10-year note. Recent record U.S. budget deficits, however, make the financing flexibility that longer-dated maturities provide critical. France's recent success in selling bonds with maturities as long as 50 years on very favorable terms (around 4%) was also surely enticing to the U.S. government. The chart below compares the long bond TLT, the intermediate bond IEF, and the shorter-term SHY.

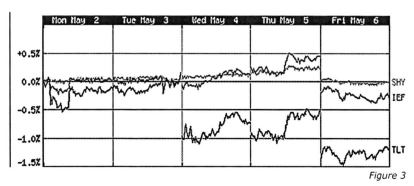

Figure 3

As the chart above shows, the potential for a new supply of longer-dated maturities sent the price of the long bond fund TLT lower Wednesday to a loss on the week. IEF was less affected by news of possible new supply.

DISCUSSION:

Although there were rumors earlier of the reintroduction of the long bond, this is big news. Over the past year, the strength of portfolios of bonds with longer maturities has been surprising. Everything about today's economy, characterized by rising oil prices, inflation threats, strong economic growth, a weak dollar, and record budget deficits, suggests higher yields for longer-dated government bonds – and lower prices for bond ETFs like TLT and IEF. Yet yields remain low and prices high. Clearly, a lot of this has to do with China's appetite for U.S. government bonds. China is buying hundreds of billions of dollars worth of 10-year notes. But part of the picture is likely the supply issue – the simple fact that no 30-year bonds were available. If the 30-year long bond is re-introduced, this may allow TLT to fall again, especially relative to IEF.

As the chart in Figure 3 above shows, bonds, especially TLT, fell on the news. But bonds were possibly also weak in the context of the ebbing fear in the market, as shown by the lower VIX in Figire 2, as well as a better outlook for equity products, as shown in Figure 1. Perhaps the combination of these three apparent negatives for bonds has not yet been fully absorbed by the market, and there is opportunity to short a bond fund like TLT on this weakness.

The question to ask here is: How much further can long-term yields fall? I think there may be an opportunity to sell TLT short.

PORTFOLIO:

The coupons due on the short TLT position can be financed with a long IEF position that should be less vulnerable to the fall in long-term bond prices I continue to anticipate. I want to combine a short TLT position with a long IEF.[1] The QQQQ is up more than 2% this week and ended at its high for the week. Add a cubes position on these new highs also.

---

1   Unfortunately, there is no TLT available to short. So I am selling five September at-the-money calls on TLT and buying five September at-the-money calls on IEF as an offset. This options trade is represented as a one-long, one-short position in the portfolio table below.

~ Is the Re-Introduction of the Long Bond Good for Investors?~

ETF	LW	TW	P/L
XLE	9	9	↓0.4%
IDU	12	12	↑0.8%
DIA	2	2	↑1.2%
QQQQ	0	1	
IEF	0	1	
TLT	0	-1	
Portfolio +0.5%			

LW: Last Week | TW: This Week | P/L: Profit/Loss

Week 46: May 9 – May 13

QUESTION:
- How to Protect the Ego When Losses Mount?

INDICATORS:
- EN, GD, CC

SITUATION:
The industrial market frustrated the bulls this week, losing ground despite lots of positive news: The March trade deficit came in better than the expected -$62 billion at a (mere) -$55 billion; April retail sales that jumped 1.4%, twice the forecast, and oil back below $50, at $48.67 a barrel Friday. The chart below compares the 1-week return of the DIA, the SPY, and the tech-heavy QQQQ.

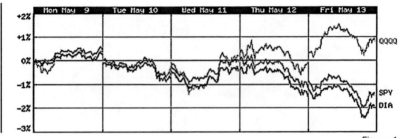

Figure 1

The chart above shows the tech-heavy cubes outperforming. As has been the case often this year, these gains are coming at the cost of the energy sector. The chart below compares the performance of ETFs in software (IGV), semiconductors (IGW), and networking (IGN) with energy sector benchmark XLE.

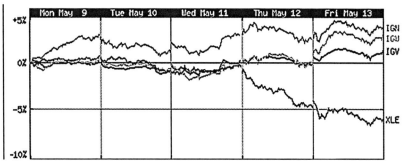

Figure 2

In another reversal of fortune, U.S. benchmarks outperformed European funds this week. On a 3-month basis, European ETFs are par-performing U.S. benchmarks, but on a 6-month, 1-year, and 2-year basis, European funds have dramatically outperformed broad-based domestic ETFs. Hurting European ETFs this week was a stronger dollar, trading to $1.2632 euros, a 7-month high and $107.3300 against the yen. Dollar bulls were helped by the smaller March trade deficit and the increase in April retail sales. The chart below compares the European EMU Index (EZU) and the Japan Index (EWJ) with the U.S. benchmark SPY.

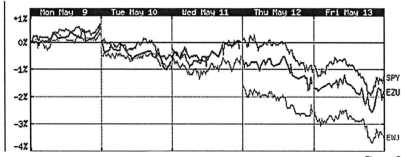

Figure 3

Fixed-income had a decent week, with longer-duration fixed-income funds outperforming the shorter-duration funds. The long bond fund TLT led. Treasury bond funds with shorter maturities, such as IEF and SHY, were higher. But LQD, the fixed-income ETF specialized in corporate bonds, lost money on the week, despite holdings with maturities similar to IEF. This may be fallout from the General Motors (GM) downgrade and perhaps a sign that investors' love affair with higher-yielding corporate bonds is ebbing. The chart below compares the weekly returns of TLT, IEF, SHY, and LQD.

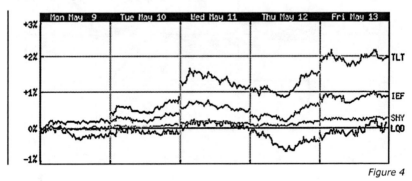

Figure 4

DISCUSSION:

Oil lower, technology higher – OK. But as for the rest – where is the pattern? Broad-market down and bonds up– on lower oil? Japan lower. Many traders try to bully the market. They would rather do anything – including lose money – than admit to being wrong. This is trading with the ego and is fatal. "What is the ultimate rationalization of a trader in a losing position? 'I'll get out when I get even.' This protects the ego."[1]

How to protect the ego? Don't protect it! The best thing to do when there are losses is think less, especially about oneself, to seek in fact to cut oneself out of the picture by automating decision-making in crisis with a consistent strategy. Whenever I have a big loss, I cut the losing positions and add to positions that are working (if any are working). Yes, I hate to sell oil here. This is a good position and has made a lot of money and is still in an uptrend. But this is what is losing now. Once again, I am in danger of falling in love with the position. My very resistance to selling is another, and in fact decisive, reason to sell.

---

1 Schwager, Jack D. *Market Wizards. Interviews with Top Traders.* New York: Collins Business, 1989. Interview with Marty Schwartz, p. 278.

PORTFOLIO:

I want to cut back on DIA and IDU. But the cubes position is working and can be increased here. The combination of adding to the cubes and decreasing oil and the DIA is a bet that the tech-oil link continues – with a gentle nudge in the opposite direction.

ETF	LW	TW	P/L
XLE	9	5	↓5.6%
IDU	12	9	↓2.0
DIA	2	1	↓1.7%
QQQQ	1	2	↑1.0
IEF	1	1	↑0.9%
TLT	-1	-1	↓2.0%
Portfolio -2.4%			

LW: Last Week | TW: This Week | P/L: Profit/Loss

Week 47: May 16 – May 20

QUESTION:
- *What Made the Biggest Rally of the Year?*

INDICATORS:
- EN, CC, GD

SITUATION:

Inflation was benign, with core CPI for April coming in at 0%. April housing starts and building permits topped expectations. Thursday's jobless claims were lower than expected. These reports helped to sustain a rally that got under way Tuesday afternoon. The chart below shows the DIA, SPY, and QQQQ.

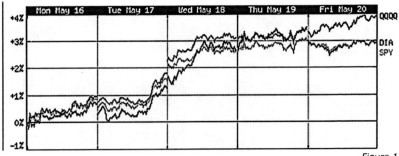

Figure 1

Market volatility, often a measure of investor fear, fell sharply starting Tuesday afternoon. The chart below shows the VIX, which measures the volatility of the S&P 500 Index.

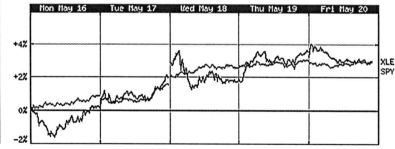

Figure 2

The charts above show that the rally started Tuesday afternoon as the VIX fell. Last week's unusual phenomenon, broad-based technology ETFs like the QQQQ outperforming the blue-chip funds like the DIA, continued this week. Is this an emerging trend?

Lower oil also helped to fuel the rally. Oil fell almost $2 a barrel, to close Friday at $46.80. Fed Chairman Greenspan boldly maintained in a speech to the Economic Club of New York that oil would not necessarily prove to be a sustained threat to the economy. Certainly, this week's lower oil didn't hurt the energy sector, which powered higher, in line with the broad market, though on greater volatility. The chart below compares energy sector fund XLE with the SPY.

Figure 3

The dollar also continued to strengthen this week, gaining again on all major currencies. Dollar appreciation was part of the story of the outperformance of U.S. equity to foreign benchmarks. The chart below compares Japan (EWJ), the Europe 350 Index (IEV), and the United Kingdom (EWU) with the SPY.

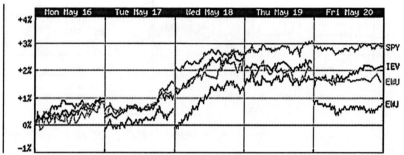

Figure 4

DISCUSSION:

The most important cause for bullishness this week is last week's selling. The market sometimes cycles. As everything was down last week, everything is up this week. Nevertheless, it is strange to see the oil and technology sectors move up in tandem. Maybe Greenspan is right, and the inverse relationship between oil and the rest of the market health is breaking down. I mistrust this. Higher oil will almost certainly mean a sustained threat to the economy.

On a psychological and technical basis, the biggest winners this week were arguably REITs. REIT ETFs are up almost 15% in seven weeks. This week, they continued their climb. The chart below compares ICF with the benchmark SPY.

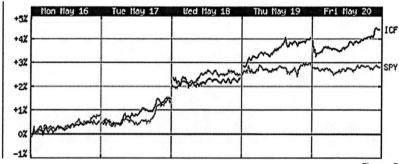

Figure 5

REITs are now positive for 2005 and trading at all-time highs. I want to get involved here on the long side, but because I have yet to make any money trading REITs and have sworn to do nothing, I'm doing nothing. Besides, REITs still look risky this year because of their debt load. The average debt to capital for REITs exceeds 50%. REIT ETF investors seem to be ignoring this, focusing instead on the high yields REITs provide.

It is hard for me to not think of REITs recent gains as a bubble. The recent sound bite is investors arguing that rising rates will make home-buying less affordable and increase the value of the rental property held by REITs. To me this sounds like double-talk: "Have your cake and eat it too." (Would they not say that falling rates are good for REITs?)

PORTFOLIO:

The portfolio strategy when things are working has been to add to the winners and cut back on the losers. XLE is working, QQQQ is working. Everything is up this week. Add modestly to everything.

ETF	LW	TW	P/L
XLE	4	7	↑2.2%
IDU	10	12	↑2.5
DIA	1	2	↑3.2%
QQQQ	2	3	↑3.9%
IEF	1	1	↑0.2%
TLT	-1	-1	↓0.7%
Portfolio +2.1%			

LW: Last Week | TW: This Week | P/L: Profit/Loss

Week 48: May 23 – May 27

QUESTION:
- *Nasdaq Up Eight Days Straight: Time to Buy?*

INDICATORS:
- EN, GD, SF, MR

SITUATION:
The broad market overcame rising oil prices, to end higher for the second week in a row. Economic news was good, with existing home sales for April in at a record 7.18 million. April durable goods orders also beat expectations, in well above the 1.3% expected, at 1.9%. The chart below shows the returns of the SPY, DIA, and QQQQ.

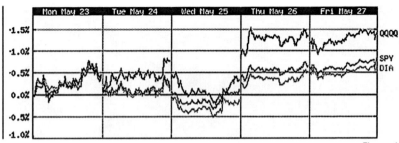

Figure 1

As the chart above shows, the tech-heavy QQQQ outpaced the

SPY and DIA. Earlier in the week, the market stagnated, as it continued to digest last week's big returns. The rally started on Thursday, when investors decided that first-quarter GDP growth at 3.5% pointed to a strong expansion on diminishing inflation risk.

Practically unheard of in the early months of 2005, this is the third week in a row the QQQQ outperformed the blue-chip benchmarks. This performance means that, for now at least, investors continue to bid up the beaten-down technology sector. This week's return was especially impressive because earlier tech strength was accompanied by lower oil prices. This week, technology managed outsized returns despite substantially higher oil.

DISCUSSION:

Technology ETFs are looking hot again. As of the close Tuesday, the tech-heavy Nasdaq QQQQ is up eight days in a row, part of a streak it has sustained since mid-April. The trend encompasses broad sectors within the technology: networking (IGN), software (IGV), and the semiconductors (IGW).

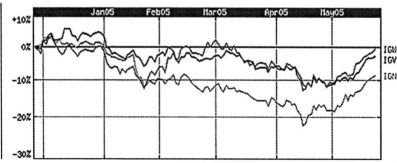

Figure 2

As the chart above shows, during the last few weeks (since about mid-April), technology has been a great place to be. Year-to-date technology is down, but it looks like the sector is close to new 52-week highs. Is this a good time to add to the tech position?

In addition to the improving technical picture, there are solid fundamental reasons for a possible improving near-term outlook for technology. Hardware demand is picking up: PC bellwethers Dell Computer (DELL) and Hewlett-Packard (HWQ) recently reported strong first-quarter results. Texas Instruments (TXN) and

Qualcomm (QCOM) report that they are expecting further increased shipments of W-CDMA handsets in the second quarter. The use of Bluetooth for wireless headsets and other wireless products also appears to be expanding. The Semiconductor International Capacity Statistics (SICAS) recently published first-quarter results for circuit wafer fabrication capacity and utilization. Wafer utilization is stable at 85%. Also promising: Historically, demand for DRAM chips picks up in the beginning of summer. DRAM prices may be nearing a bottom.

The one single biggest and oft-repeated problem with the tech sector is continued lackluster corporate spending. After two years of growth in the mid-teens, corporate IT spending for 2005 looks as if it will be mired in the single-digit growth range. Longhorn, the next-generation operating system from Microsoft (MSFT), anticipated as a driver for corporate hardware and software spending, will not be available before 2006. There appear to be few near-term drivers for a growth in corporate tech spending through the end of 2005.

High oil prices have tended to harm the performance of broad-market ETFs generally, and the historically high betas of technology ETFs make this effect more important. For example, during the last six weeks, as technology has been recovering, energy ETFs have been in a downtrend. The chart below compares the 1-month returns of XLE and QQQQ.

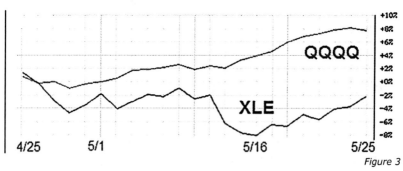

Figure 3

Although the tech sector has been strong in recent weeks, current technology strength needs confirmation and additional catalysts. It is still too early to say whether or not this recent move implies a fundamental shift in the direction of technology. In the meantime, I want to continue to add to strength in the cubes.

PORTFOLIO:

Oil closed Friday at $51.85 a barrel. Oil benefited from the usual (but overdue this year) seasonal speculation that summer driving season would increase demand, as well as the perception that an increasingly healthy economy would help the travel business and drive up demand for jet fuel. On top of this, some oil bulls were motivated by reports Friday that Saudi Arabia's King Fahd – a major stabilizing force in the oil market – has been hospitalized. The chart below shows the 1-week return of the domestically focused XLE and the internationally focused IXC.

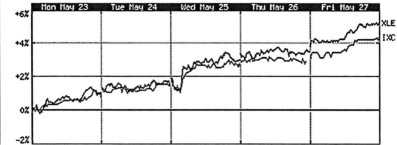

Figure 4

As the gentle chart above shows, both funds were up a steady 4% on the week.

ETF	LW	TW	P/L
XLE	7	9	↑5.9%
IDU	12	12	↑0.8%
DIA	2	2	↑0.7%
QQQQ	2	3	↑1.5%
IEF	1	1	↑0.5%
TLT	-1	-1	↓0.4%
Portfolio +1.5%			

LW: Last Week | TW: This Week | P/L: Profit/Loss

Week 49: May 31- June 3

QUESTION:
■ *Summer Is Here: Time to Sell Volatility?*

INDICATORS:
■ GD, CC

SITUATION:
School is out for summer and the market is thinning out. Investors had to wait until Friday for the big news this week: the May employment report. It disappointed. The government reported the slowest job growth in two years. Nonfarm payrolls, expected to be 175,000, missed by a mile, coming in at 78,000.

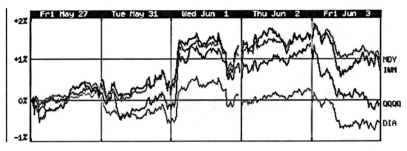

Figure 1

The chart above compares the QQQQ with the small-cap fund IWM, mid-cap fund MDY, and the large-cap DIA. Most small-cap and mid-cap ETFs actually finished the week to the plus side, outperforming large-cap funds, but overall, the market was flat.

A strengthening dollar helped domestic ETFs best most foreign funds, but foreign performance also lacked cohesion. The chart below shows three foreign benchmarks: EAFE, Index fund EFA, Japan's EWJ, and Europe's EMU Index (EZU).

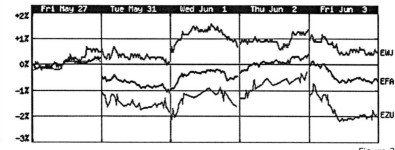

Figure 2

DISCUSSION:

Summer is typically a slow time for the market. Money managers go on vacation. Market volume disappears. Summer can be a hard time to make money on a trade. The popular expression "sell in May and go away," captures this sentiment. The first rule of summer trading is not to be fooled by a change in the mood of the market.

▲ *The best answer to the question "Will the market go up or down?" is usually: neither. It will remain unchanged. Especially true in summer.*

Recall the summer of last year, when many key broad-market ETFs frustrated both the bulls and the bears, ending summer almost precisely as they began it. The table below shows the returns of several key broad-market ETFs for the summer of 2004:

	June 1	August 31	Summer Change
DIA	101.91	101.89	0%
SPY	112.46	111.11	-1.2%
EWJ	9.98	9.98	0%
EZU	59.82	59.85	0%
EFA	138.66	138.53	0%
QQQQ	36.27	34.02	-6.2%

Table 10

With the exception of QQQQ, which lost 6.2% over the summer

months, these other key ETFs literally did nothing last summer.

The table below compares the summer returns (June 1 to August 31) for the six ETFs in the table above. In addition to those ETFs listed above, the table below shows some of the most popular ETFs of this year: the small-cap IWM, the mid-cap MDY, oil, XLE, utility sector XLU, and the long bond fund TLT.

	2000	2001	2002	2003	2004
DIA	6.1%	8.8%	-12.9%	5.6%	0%
SPY	6.0%	9.6%	-14.3%	4.0%	-1.2%
EWJ	-3.1%	-16.0%	-13.6%	18.9%	0%
EZU	n/a	-8.7%	-15.8%	13.8%	0%
EFA	n/a	n/a	-13.7%	5.8%	0%
QQQQ	19.3%	-19.6%	-21.7%	12.4%	-6.2%
IWM	10.7%	-5.9%	-19.5%	11.8%	4.1%
MDY	12.6%	-5.6%	-16.4%	8.1%	-2.8%
XLE	0.4%	-14.2%	-15.9%	0%	6.4%
XLU	1.5%	-9.0%	-17.9%	-5.6%	6.4%
TLT	n/a	n/a	n/a	-10.9%	6.6%

Table 11

On the basis of this small (5-year) sample size, at least in terms of its symmetry, last summer looks anomalous. Most ETFs showed big swings for the summer months. Does this mean that selling volatility would have been a bad trade for those summers? In 2002, market volatility, as measured by the VIX, increased from about 23% on June 1 to about 33% by August 31, peaking in earnings season. The chart below shows this increase in volatility.

Figure 3

The chart above shows the increase in volatility in the summer of 2002. The smooth lines represent the 20-day and 30-day moving average. With the exception of 2002, every other year of the last five saw volatility fall over summer. For example, the chart below shows the fall in volatility in the summer of 2000.

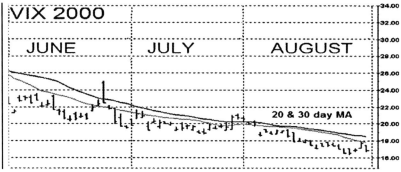

Figure 4

Here again, the smooth lines represent the 20-day and 30-day moving averages. As the chart shows, although there was a June peak, overall volatility fell from about 21% in the beginning of June to near 18% by the end of August.

Because the VIX is beginning June of this year near a multi-year low of 12%, it looks as if selling volatility does not have the same upside potential of previous summers. Making decent money selling volatility this summer is a waiting game – wait for a panic, a spike in volatility, like the spikes in late July and early August of 2002, and then sell that spike. There will probably be an opportunity to do this at some point during summer.

Does the current low volatility mean that this is a good time to buy volatility? Probably not. The problem with a long volatility strategy is that it is always very difficult to time. While this summer will likely see an increase in volatility above the current 12% at some point, the question clearly is: when? It is unlikely that there will be anything rivaling the summer of 2002, when broad-market ETFs were trading to new lows every day.

In addition to the historical uptick in volatility in late July, crude prices will be a factor. If the price of crude makes significant new highs, fear may return to the marketplace and an investor in volatility at 12% may be rewarded

DIA and QQQQ have halted their upward ascent. Both are negative this week. Oil is up, tech down. This is more familiar territory. No change.

PORTFOLIO:

Bonds are looking toppy. Historically, 10-year yields below 4% have not been sustainable. The benchmark 10-year bond yield moved to new 52-week lows, below the psychologically important 4% level. The chart below shows the yield on the 10-year Treasury note, represented on the *x*-axis. As the chart shows, yields are now lower than any time over the last year.

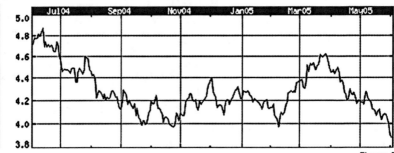

Figure 5

So the fixed-income hedge lost money again this week, and yet it looks better and better the lower the yield falls. Given the high oil price and the potential for inflation, it doesn't seem likely that yields can keep falling.

ETF	LW	TW	P/L
XLE	9	9	↑1.5%
IDU	12	12	↑1.3%
DIA	2	2	↓0.6%
QQQQ	3	3	↓0.2%
IEF	1	1	↑05%
TLT	-1	-1	↓1.9%
Portfolio +0.8%			

LW: Last Week | TW: This Week | P/L: Profit/Loss

Week 50: June 6-June 10

QUESTION:
- Why Are Treasury Yields So Low?

INDICATORS:
- EN, SF, CC, TA

SITUATION:
As volume continued to drain away from summer markets, most sectors ended the week little changed. One exception was technology. After several weeks of outperforming, the broad-market technology fell on weakness in software and networking. The tech-heavy QQQQ underperformed the SPY and DIA.

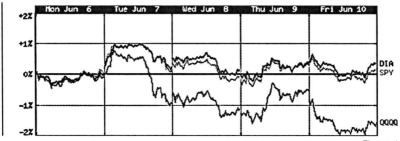

Figure 1

Bonds moved up early in the week, as investors speculated that the Fed's rate cut plan was nearing an end. But on Friday, bonds were crushed as Greenspan refused to provide any guidance for a change in the Fed stance, a move that ended speculation that a pause in interest-rate increases was imminent. The chart below shows the returns of the long bond fund TLT, intermediate-term fund IEF and the near-term fund SHY.

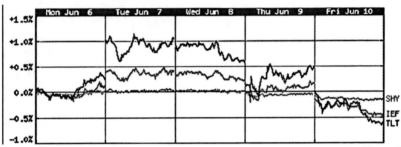

Figure 2

Despite the Fed chairman's recent remarks, bonds, especially the longer-dated TLT, have remained strong. Below is a 1-year chart of the long bond fund TLT, intermediate-bond fund IEF, and near-term fund SHY.

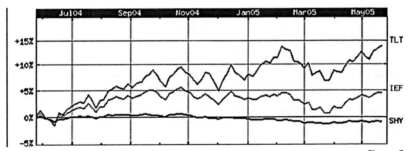

Figure 3

As the chart above shows, ETFs holding shorter maturities, such as SHY, have not done as well as portfolios of longer maturity treasury notes such as TLT and IEF. The strong performance of TLT and IEF has baffled many market watchers. Today's economy, characterized by rising oil prices, inflation threats, strong economic growth, a weak dollar, and record budget deficits, suggests higher yields for longer-dated government bonds – and lower prices for bond ETFs like TLT and IEF. Yet yields have remained low.

One of the people confused by the falling yield is Fed chair Alan

Greenspan. In mid-February at the Federal Reserve Board's semiannual monetary policy report to Congress, Greenspan referred to low yields (and high prices) of treasuries as a "conundrum". What puzzled Greenspan back in February was that, although the Fed had at that point increased rates by 150 basis points (1.5%), yields on the 10-year Treasury note had not gone up at all, and had, in fact, fallen.

What might explain the low long-term Treasury yields?

DISCUSSION:

The simple answer is demand for U.S. Treasury debt. Who's buying? Foreign Central Banks, especially in Asia, are buying Treasuries to prevent the appreciation of the dollar. Currency weak in relation to the dollar makes goods cheaper in dollar terms for the American consumer. Cheaper goods in turn keep U.S. inflation low. But the more the U.S. consumer buys from these countries, the more dollars they receive, and the more money needs to be recirculated into Treasuries. This means that low Treasury yields are tied to U.S. consumption.

Low inflation is also helping to keep Treasury yields low. Here, Greenspan also plays a role. Greenspan has said that he will leave the Fed in January of next year. The markets believe that, in his final six months, Greenspan will prove himself to be a banker at heart – and all bankers fear inflation above all.

In addition, there are the more subtle (and more dubious) arguments for permanently lower yields. They include the increasing popularity of floating rate debt, both in the mortgage and corporate lending markets, as well as structural changes in how large pension plans are funded. Historically, floating rate debt increases the demand for long-term Treasuries. Some experts claim that, for accounting purposes, large pension plans are moving into treasuries.

The yield on the 10-year Treasury note, after rising sharply on Greenspan's remarks in February and moving higher in February and March, is now back down to its mid-February level, when the Fed chairman first expressed his surprise.

The chart below compares the current yield curve to the historical 6-month and 1-year yield curve.

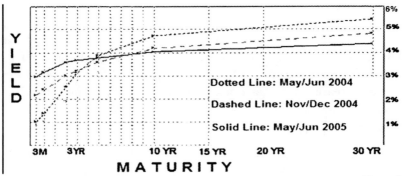

Figure 4

The chart shows three yield curves. The curve in solid black is current, the curve from six months ago is represented with a dashed line, and the curve from a year ago is dotted. The chart shows that the yield curve has been flattening. A year ago, the 3-month yield (shown at the far left of the graph in dotted blue) was at 1%. The 30-year yield (shown at the far right in dotted blue) was 5.5%. The spread between the two was 4.5%, or 450 basis points. As the black line shows, today's spread is just 1.5%, or 150 basis points. Over the last year, in other words, the yield curve has flattened end-to-end almost 3% or 300 basis points. Most of the slope still in the curve is in the treasuries with 1- to 3-year maturities.

To put this into context, a yield of 4% means that a bond investor receives less than 1% more compensation for lending the government money for ten years through the purchase of a Treasury note, as compared to the shorter term 1-year or 2-year Treasury bill.

The Fed is probably not happy with these low yields. The reason is that mortgage rates are determined on the basis of the 10-year note. So, low rate financing continues to be available, and property values remain stubbornly high. With yields so low, the Fed has not been successful in letting the air out of a potential bubble in the real estate market.

Of course, the Fed has a problem: It cannot talk up yields. When a consumer takes advantage of low rates by taking out a home equity loan and buys with that money goods made in China, he perpetuates the cycle, because the Central Bank of China then turns around and buys Treasury notes, which, by further driving down yields, encourages the consumer to borrow more money, and so on.

The Fed's solution seems to be further rate hikes. If the yield on the 10-year Treasury remains where it is today and the Fed continues to raise the short-term rate, the 1% spread between yields on the 2-year and yields on the 10-year will narrow further.

PORTFOLIO:

No one is yet talking about an inverted yield curve (when short term rates are higher than long-term rates). Barring the possibility of an inverted curve, things don't look as if they can get much better for long bond ETFs; TLT looks fully valued. I'll remain short TLT.

There is talk about hurricane season again and the possibility that bad weather on the Gulf Coast may lead to supply problems if shippers and refiners are forced off-line. On Friday, several large producers evacuated nonessential personnel from facilities, as the first tropical storm of the season, Arlene, made its approach. Though the partial evacuation in this case is more psychological than material, with oil in tight supply, it is a reminder of the likely volatility in the oil contract in the months ahead. With hurricane season coming, there seems no reason to sell. New highs on oil? Time to add more. The chart below compares this week's returns of XLE and IXC with the benchmark SPY.

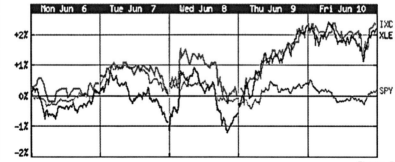

Figure 5

Add to oil on this strength.

Technology, as shown in Figure 1 above, is crumbling. I'll cut back the QQQQ position.

~Sector Trading: A Year in Exchange Traded Funds~

ETF	LW	TW	P/L
XLE	9	12	↑2.4%
IDU	12	12	↑1.1%
DIA	2	2	↑0.4%
QQQQ	3	1	↓1.4%
IEF	1	1	↓0.3%
TLT	-1	-1	↑0.4%
Portfolio +1.0%			

LW: Last Week | TW: This Week | P/L: Profit/Loss

Week 51: June 13- June 17

THEME:
- *Is Alternative Energy a Viable Oil Play?*

INDICATORS:
- EN, GD, SF

SITUATION:
Oil ripped up almost 10% this week, to close Friday at an all-time high, at $58.47 a barrel. The big move in oil came despite an OPEC promise to raise its daily output quota by 500,000 barrels, to 28 million barrels.

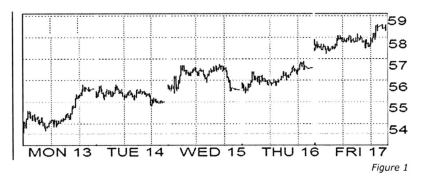

Figure 1

The chart above, with the dollar price per barrel on the right axis, shows this week's increase in the price of crude.

In movement typical of the last two years, energy ETFs followed crude higher. The smooth chart below, comparing the domestically focused XLE with the more globally focused IXC, shows that energy ETFs were once again an easy trade this week.

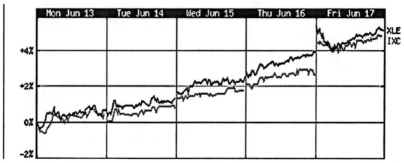

Figure 2

Why did oil go up so much? First, investors are worried that good intentions from Saudi Arabia may be becoming increasing irrelevant if world demand continues to grow. This is very bullish for oil. Second, seasonal concerns may also be a factor here, as investors ponder supply disruptions during hurricane season. Third, there may be production capacity problems. Refineries' struggle to keep up with record demand may further complicate supply.

DISCUSSION:

Will alternative energy benefit from these supply problems? The chart below compares the recently introduced PS Wilderhill Clear Energy ETF (PBW) with the XLE and SPY.

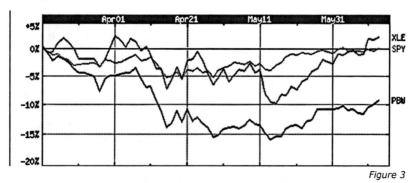

Figure 3

As the chart above shows, on a 7-day basis, PBW and XLE track each other closely. Both hit their 3-month lows in mid-May, with XLE down over 10% and PBW down 15%. While XLE has recovered since then to trade higher than the benchmark SPY, PBW continues to lag.

PBW tracks the Wilderhill Energy Index, known as ECO. A look at the ECO index is useful for assessing the direction of PBW. The 2-year chart below compares the ECO index with the price of a barrel of crude oil. On the lefthand axis is the Wilderhill Index. The righthand axis shows the price of a barrel of crude.

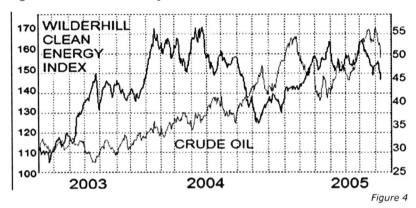

Figure 4

Over the 2-year period represented in the chart, the Wilderhill Index correlates with the price of crude, though on greater volatility. However, while crude in the past two years has nearly doubled – from about $30 a barrel to above $55 – the Wilderhill index is up by just a third.

Clearly, there are problems with using the price of crude, or companies processing crude, as an indicator for renewable energy companies such as those held in PBW. The most obvious problem is that renewable energy companies primarily target electricity generation, effectively competing with utilities, not oil companies. According to the Department of Energy, oil provides only 3% of electricity production. Compare this number to that of coal (53%), nuclear (21%), natural gas (15%) and hydropower (7%). Although electricity prices have moved higher in the last two years, as commodities from coal to uranium have soared, the prices have not moved as quickly as oil. The Wilderhill Index and PBW seem to follow the price of crude more closely than the steady upward tilt of the utilities ETFs.

PORTFOLIO:

It's been pretty easy to make money in the energy and utilities sector over the last year, simply by buying an oil sector fund XLE or a utilities fund like IDU. Although PBW has underperformed in the last three months, it looks as if it will ultimately follow oil higher. PBW also potentially has a kind of sweet spot – the buzz of tech-sector growth and the momentum of the oil sector. Increase the oil position and add PBW.

ETF	LW	TW	P/L
XLE	12	15	↑5.3%
IDU	12	12	↑0.5%
DIA	2	2	↑1.0%
QQQQ	1	1	↑1.0%
IEF	1	1	↓0.2%
TLT	-1	-1	↑0.6%
PBW	0	1	
Portfolio +1.8%			

LW: Last Week | TW: This Week | P/L: Profit/Loss

Week 52: June 20 - June 24

THEME:
- *What Are Fundamental Drivers of Oil and Utilities?*

INDICATORS:
- EN, FA, CC

SITUATION:
Fear has returned to the market. Investors are selling equities and buying bonds. The chart below shows the mirror-like return of the benchmark equity ETFs DIA and SPY with the long-bond fund TLT.

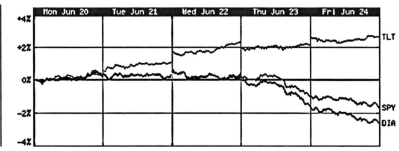

Figure 1

What accounts for the divergence? The fall in the broad market coincides with the price of crude passing $60 a barrel for the first time Thursday. The chart suggests that investors responded by

~225~

selling equity and fleeing to safety in bonds. By Friday, yields on the benchmark 10-year note fell back below 4% to 3.914%.

Other than bonds, there were few bright spots domestically. The utilities sector is one area that remains untarnished. The chart below compares this week's returns of the three utility ETFs: IDU, XLU, and VPU.

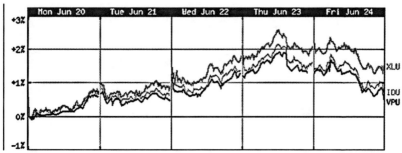

Figure 2

As the chart above shows, the big run-up in utilities started Wednesday. They may have got a boost from President Bush's remarks that the United States should return to building nuclear power plants. Although nuclear power typically faces opposition, as the president noted, Americans are consuming electricity faster than they are producing it.

Internationally, Asian ETFs outperformed and Latin American underperformed broad U.S. benchmarks. The first chart below shows the performance of Asia, with China (FXI) and Hong Kong (EWH) leading, and Japan (EWJ) lagging.

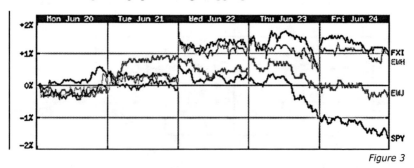

Figure 3

As the chart below shows, Brazil (EWZ) and Mexico (EWW) also suffered:

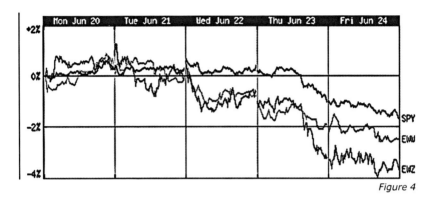

Figure 4

The two charts show that foreign markets are deeply linked to the performance of the United States markets. In this case, the sell-off in the United States late Thursday sparked a worldwide sell-off, with Brazil and Mexico dropping further than the less closely linked Asian economies.

DISCUSSION:

Oil ETFs have been strong for the last 52 weeks and continue to be strong, both technically and fundamentally. There is no sign yet at all that it is time to exit a long oil position. But on a longer-term basis, the fundamental story for oil is deteriorating. Psychologically, one sign of this is the increasing circulation of the story that it is expensive because "There is only so much oil in the world before it all runs out." In the way that there is only so much gold or silver, or beach-front property, while basically true, this is fundamentally a misunderstanding.

The oil price is not high because the world is running out of oil. Oil is high because there is not enough immediate supply, which means not enough supply today, and because not enough oil is being pumped from the ground. Until there is more supply (meaning more oil is pumped), oil will be expensive. In the short term, with virtually all of production operating at capacity, there is real risk of supply-side shock, which, if it happens, will drive the oil price still higher.

In the longer term, like gold, soybeans, wheat, or sugar, oil is a commodity. High commodity prices attract new supply. While the Dow Jones Industrial Average rose over 10,000% in the past century, the prices of wheat, sugar, soybeans and other commodities have actually fallen. Investing in oil companies or energy ETFs, which are portfolios of oil companies, is a good way to profit from higher oil prices without being stuck owning a commodity.

Utilities continue to be more than strong, unassailable. The chart for utilities continues to be beautiful. Unlike the growth sectors of yesteryear, such as technology and biotech, the upward movement of utilities is steady and predictable. IDU currently has a 3-month beta of 0.69, among the lowest of any ETF, so utilities investors have had the best of all possible worlds: valuation growth on low volatility.

But why are valuations in utility ETFs climbing? Have growth prospects for the utilities sector increased? Is the utility sector newly capable of rapid growth? The answer is no. Although electricity prices have moved higher in the last two years, as commodities from coal to uranium have soared and demand for electricity has increased, growth has been steady but not spectacular.

The most important factors influencing the popularity of utility ETFs are the negative general outlook on the market generally and the sturdy promise of cash dividends from utilities. Dividends can be thought of in terms of yield, which is the relation of the price of the ETF to the dividend it pays. As with bonds or any dividend-paying security, as the price goes up, its yield falls. Some utilities are increasing dividends. This means that they are paying out more cash than they were before. This has helped to prop up yields but not enough to offset the fall in yields due to the rapid increase in valuation. In other words, even after increasing dividends, the rapid rise in valuation has meant that the yield on utilities has fallen and continues to fall. As bond yields rise, the dividends utilities pay will look less attractive, and utilities will come under pressure. But the chart will likely waver at this height and utilities will become more volatile.

As of yet, there remains little sign of this happening. For now, keep to the system. Hold utilities for next week, cut the oil ETF position slightly on this decline. Exit the broad-market SPY and QQQQ positions.

PORTFOLIO:

ETF	LW	TW	P/L
XLE	15	12	↓2.0%
IDU	12	12	↑0.5%
DIA	2	Closed	↓3.0%
QQQQ	2	Closed	↓2.4%
IEF	1	1	↑1.1%
TLT	-1	-1	↓2.2%
PBW	1	1	↓0.1%
Portfolio -0.8%			

LW: Last Week | TW: This Week | P/L: Profit/Loss

~ End of Year ~

In Conclusion

The trading strategy followed over the past 52 weeks involves buying on strength, selling on weakness, and holding when there is no movement. Specifically, this was accomplished by rigorously adding to positions when they advanced more than 1% on a weekly basis, cutting back on positions when they declined more than 1% on a weekly basis, and holding positions advancing or declining less than 1% on a weekly basis. In the case of particularly volatile ETFs, such as iShares Nasdaq Biotechnology (IBB), the 1% band was relaxed. In special situations, when a significant gain or loss in the portfolio (usually two standard deviations) was accompanied by an unusually large move in the market, contrarian strategies, such as exiting a profitable position or adding to a losing position, became options. In these rare situations, seasonal factors and broad market rhythms became especially important.

Portfolio Results

Portfolio returns over the past year were 16 down weeks, with an average weekly loss of 1.3%, and 35 up weeks, with an average gain of 1.4%. The overall average weekly return was 0.56% compared to 0.17% on the S&P. Although the portfolio performed well compared to this benchmark, a simple long position in an energy ETF such as the Energy Select Sector SPDR (XLE) would have done better. XLE returned 43%, or 0.82% on a weekly basis, during this period. The chart below shows the portfolio's gain and loss.

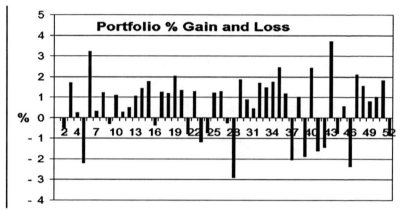

Figure 1

The dark vertical bars in the chart above show each week's percentage return as enumerated in the lefthand scale. The chart shows the profit or loss on a percentage basis only. The percentage chart above does not indicate numeric profit and loss. The amount of money invested in the portfolio fluctuated on a week-to-week basis, increasing over the course of the year. The chart below shows the portfolio's numeric or dollar gain and loss on a weekly basis.

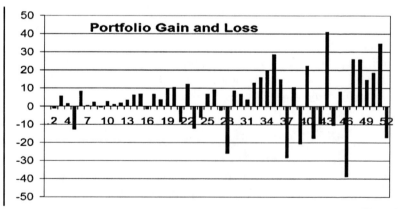

Figure 2

~ *In Conclusion* ~

As the above chart shows, portfolio gains and losses in the second half of the year are more important than in the first half, when relatively less total money was invested. Due to position sizing, both gains and losses became more dramatic in the later weeks as the overall size of the portfolio increased. As the chart above shows, in single best and worst weeks for the portfolio, week 43 and week 46 respectively, the portfolio gained and then lost more in real terms than it had during the entire first quarter.

Through a combination of position sizing and active trading, the portfolio grew to over ten times its initial size over the course of the year. Most of this size increase came from position sizing, rather than market gains. The chart below shows the overall growth of the portfolio.

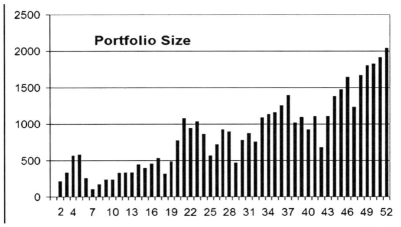

Figure 3

As the chart shows, the portfolio suffered periodic losses and was cut back several times over the year. It was downsized dramatically, for example, on week 28, after the market was routed in its first weekly session of 2005. It also was cut severely on week 42 after a week of selling leading up to April 15, 2005, tax day. The portfolio fell sharply on week 46, beginning May 9, when a sharp sell-off in oil accompanied a technology advance. This was especially costly because the portfolio was mostly long oil and short tech. One of the consequences of building a momentum portfolio and evaluating picks on a week-to-week basis is that the amount invested can see significant fluctuation. According to the momentum strategy followed, in weeks when the portfolio loses money it is also cut back in size, and the total amount invested drops sharply. Portfolio size fluctuation reflects the strategy of adding to winning positions and cutting back on losing positions

A Time Horizon for ETFs

Part of any trading methodology is the determination of a relevant time-horizon for market analysis and decision-making. There are hourly trends, daily and weekly trends, and trends that maintain over many years. What looks like a buy on a very short-term basis may look like a sell on a long-term basis and vice versa. Although a weekly time frame is probably too short to assess the trends in individual stocks, the book shows that for sector trading with ETFs, the week is a valid time frame and unit of measurement.

ETFs are comparatively less volatile, slower moving, than individual stocks. The diversity of their holdings makes the moves they do make generally more reliable. A week-long time frame has many practical advantages for traders. One advantage is that there is a natural period after each week (the weekend) during which market participants are forced to evaluate the previous week's action and excesses. A second advantage is that focusing on a one-week time horizon provides the trader with a practical benchmark that allows him to standardize his thinking and develop an intuitive sensitivity to market movement. A third advantage is that psychologically and culturally, we are accustomed to thinking about a week as having a beginning and an end. This mindset has implications for market behavior, and short-term trends can come and go during a one-week time horizon.

Develop a Trading Plan

There are three steps to successful sector trading.

1. Identify a trend.
2. Develop a trading strategy (for participation in that trend).
3. Execution and risk management (the actual

mechanics of participation).

1. *Identify a trend.*

To identify a trend means to identify a sector ETF demonstrating consistent price movement. Doing this requires looking closely at the historical movement of a sector. Most of the portfolio gains this past twelve months came from buying the market-leading oil and utilities sectors, and selling short tech and biotech. It is not possible to catch the whole move, because it is not clear that these trends actually exist until they are already under way. The trend for higher oil was in place at least six months to a year before the portfolio began trading. It remains in place today. Utilities, after the period of post-Enron lows, also began a multiyear upward progression. This trend maintains today. Technology, after the huge bounce in 2003, drifted lower starting in 2004. These three trends remained consistent over this period. All these trends will end, and quite possibly near simultaneously, as technology often vies with energy and more defensive sectors for investment dollars.

Fundamentally helpful to a momentum-based trend-following strategy is reliable political and monetary infrastructure. The trend of higher oil and higher utilities was held in place partly by consistent monetary policy from Greenspan and the Federal Reserve, as well as unvarying foreign policy and regulatory behavior from the White House. Because the price of oil and gas is very political, any political change can mean a huge shift in energy policy and pricing. The perception, which the whole market gradually absorbed, that the Fed would raise rates only 25 basis points per meeting helped to provide a consistency and reliability for large market trends.

Is there an alternative to trading the trend? Yes: forecasting and prediction. This involves a trader betting that he knows where a market will go. Most of the more significant losses in the portfolio over the course of this year involved unwittingly abandoning trend-following strategies and engaging in prediction.

2. *Develop a Trading Strategy*

No trader or market analyst can know where the market will go, when the market will follow a described trend, and when it will stop following that trend and merely consolidate, or even begin a counter-trend. For this reason it is important to formulate a strategy for catching the trend and for managing risk by cutting back on position size and exiting a position when a trade is not successful.

When formulating a trading strategy, backtesting is essential and will provide a guide for how to trade the trend. Backtesting involves reviewing the historical micro-movement of the specific sector to be traded. Although, of course, past results are never a guarantee of future performance, in fact, a trader can learn a lot from looking closely at how a sector or an ETF has traded historically.

A trader should look for sectors that seem to trade according to a recognizable pattern, and familiarize himself with how and when a sector moves, and how and when a sector ETF responds to news and information. Different sectors respond differently, and the same sectors often respond differently at different times. In the case of the portfolio described here, buying highs and selling lows worked well for oil service ETFs, utility ETFs, and to a lesser extent technology ETFs. It worked less well for the REITs, as week-to-week they often lacked significant follow-through. As the oil bull gets older, it will likely cease to reward the trader who buys according to this methodology, but may respond well to different accumulation strategies. The better sense a trader gains of these differences, the more successful his trading will be.

3. Execution and Risk Management

Once a trend is identified and a strategy developed and backtested, it is important to be disciplined in trade execution. Undisciplined traders always lose money because sooner or later they cease to follow trends and start trying to predict the future.

Actual trade mechanics usually work best when they are consistent with the strategy and historical patterns. But this alone is not enough. A trading style should also be comfortable to the trader. If it is not comfortable, then maintaining discipline becomes difficult or impossible. In my case, I like to be long and to add to a trending position on strength. I like to sell the position on weakness.

There are other styles of trading a trend, such as buying on dips. Although this also attempts to exploit an uptrend, it provides opposite buy and sell signals from the strategy followed here. In any case, when an ETF does not trade according to a style comfortable to the trader, there is no opportunity for him and lots of risk. It is important in this situation to stop trading this sector.

What's Ahead?

With Bush in office through 2008, there is little reason to think that there will be any surprises from Washington in the next few years. However, after serving as Fed chairman for 19 years, Alan Greenspan is set to retire in early 2006. This may create a period of uncertainty, which will influence market direction, particularly for fixed-income but also for the broad market. In some ways, this is an unfortunate time for Greenspan to be departing. Over the past year and a half, the yield curve has been flattening, if Greenspan continues on his course of raising the Fed Funds Rate 25 basis points per meeting, when he departs, the short-term rate will be at 4.5%, higher than the current long-term yield. If long-term yields do not rise, the yield curve invert. An inverted curve is widely interpreted to be an early sign of recession. Both because of Greenspan's departure and the possibility of a recession, early 2006 will be a good time to evaluate the durability of existing trends, and to look for profitable new trends.

It is also evident that the Fed, though still enormously powerful, has been increasingly undermined by two forces: the profligate spending of the White House and Congress, and the increasing impact of globalization on the domestic economy. These new pressures create forms of domestic and global instability that ultimately have muted the importance of the Federal Reserve's monetary policy decisions.

The benchmark 10-year Treasury note, which determines the cost of consumer credit and mortgages, used to be very sensitive to the shorter discount rate set by the Fed. In the last few years, the benchmark 10-year note has arguably responded more to Chinese and other foreign government purchases of U.S. government debt than to the Fed's decisions on short-term interest rates. While

Greenspan has raised rates ten times since June 2004, yields on the 10-year Treasury note did not go up, and in fact fell over this period. This new reality, with U.S. domestic monetary policy possibly determined as much by decisions overseas as by the Federal Reserve Board, will have significant ramifications in the years ahead. This is perhaps the most important new macro-economic global trend that emerged over the past twelve months.

The portfolio covered in the course of this book has focused primarily on the U.S. economy and trading sector ETFs within the U.S. context. By far the most important non-domestic ETF in terms of volume is the iShares MSCI Japan Index (EWJ), which alone trades more than twice the volume of every other non-domestic ETF combined. Still, trading volume in EWJ is less than a fourth that of the largest ETFs, SPY and QQQQ. In the coming decade, learning to trade ETFs of sovereign nations, learning to arbitrage Europe, Japan, and the United States, as well as China, Hong Kong, etc., and developing enhanced indexing strategies using these funds will provide massive new opportunities. This, I believe, is an important new direction of ETF sector trading in the years to come.

Appendices

Glossary

Arbitrage. Making simultaneous purchases in one market or security and sales in another highly related market or security. A trading strategy designed to profit from differences in markets rather than market direction.

Bear. An investor who believes the market will fall.

Beta. A measure of risk. The volatility of a security compared to overall market volatility. Market beta is 1. A security with a high beta of 1.5, for example, is expected to be 50% more volatile than the market. A security with a low beta, of 0.5 for example, is expected to be 50% less volatile than the general market.

Breakout. When an ETF moves to a new high or new low.

Broad market ETF. An ETF designed to reflect the movement of the entire market rather than one specific sector or area. The Diamonds Trust (DIA) and the Standard and Poor's Depositary Receipts (SPY) are examples of broad market ETFs.

Bull. An investor who believes that the market will rise.

Contrarian. An investor who trades opposite to the trend.

CPI. Consumer Price Index. The CPI measures the price of goods and services as purchased by the consumer. The CPI is widely used as an indicator for inflation.

Credit spread. The spread between Treasury securities and corporate debt or other non-Treasury paper.

Daytrade. A trade held for a short period. Typically, a trade closed before the end of the day it was initiated.

DOT limit. Short for Designated Order Turnaround, a position and capital limit set by the risk management staff of a trading organization.

Downtick. When the last quoted price of a security is lower than the penultimate quoted price.

Duration. A measure of the price sensitivity of a bond to an interest rate change of 100 basis points. Bonds with higher duration have greater sensitivity to interest rate changes.

EMA. Exponential Moving Average.

ETF. Exchange Traded Fund. A security that tracks an index, but

~ Glossary ~

trades like a single stock. Like mutual funds, ETFs hold a basket of companies, but unlike mutual funds, they are not actively managed and do not have a daily NAV (Net Asset Value) calculation.

Fed Funds Rate. The interest rate set by the Federal Reserve.

Federal Reserve (Fed). Government entity concerned with setting monetary policy. The Fed seeks to regulate the economy by raising and lowering interest rates.

Fixed income. A fixed income security is characterized by fixed payments and the return of principal at maturity. A bond is a fixed income security.

FOMC. Federal Open Market Committee.

Fundamental analysis. The practice of estimating the price of a security based on its inherent (rather than market) value. Fundamental analysis involves the use of economic data and an assessment of market conditions.

Hedge. A position used to offset risk.

Large-cap. Companies with market capitalizations of greater than $10 billion.

Leverage. The use of borrowing to increase the size of the position. A trade is leveraged when the dollar amount of the position is greater than the capital employed in the position.

Liquidity. A measure of market efficiency, and the extent to which an asset can be simultaneously be bought and sold without adversely affecting its price. 100 shares of IBM, for example, are typically much more liquid than, say a house or a car, which are relatively costly to buy and sell.

Long. A long position in a security is created by buying that security.

Maturity. The length of time before the principal of a bond is repaid.

Mid-cap. Companies with a market capitalization of $2 billion – $10 billion.

Moving average. A calculation of average prices

OPEC. Organization of Petroleum Exporting Countries.

Portfolio. The total assets held by an investor.

Producer Price Index (PPI). The PPI measures the price of goods at a wholesale level. The PPI includes the cost of raw goods that are used in production. The PPI is often watched for signs of inflation.

Sector trading. Trading securities based on their sector or market.

Shakeout. When investors exit positions due to bad news or adverse technical action.

Short-selling. Selling a borrowed security with the expectation that the price and value of that security will fall.

Small-cap. Typically companies with market capitalizations of less than $2 billion.

Spread. The difference between the bid and offer prices of a security.

Technical analysis. The practice of evaluating the price of a security based not on the business, management or any other factor inherent to the company but rather on its movement in the market, its chart, volume and other indicators.

Thick. Heavy volume.

Thin. Light volume.

Uptick. When the last quoted price of a security is higher than the penultimate quoted price.

Volatility. A measure of price variation and market fluctuation. A volatile market has a lot of price variability and fluctuates widely.

Volume. A measure of the number of shares traded.

Yield curve. A curve that shows the interest rate paid on securities that have equal risk over different maturities.

List of ETFs

AGG. The iShares Lehman Aggregate Bond Fund. AGG holds primarily a mix of corporate and treasury bonds. AGG tracks the Lehman Brothers Bond Index.

DIA. DIAMONDS Trust. The DIA tracks the performance of the Dow Jones Industrial Average (DJIA).

DVY. The Dow Jones Select Dividend Index Fund. DVY holds stocks with strong dividend-per-share growth. It corresponds to the Dow Jones Select Dividend Index.

EEM. The iShares MSCI Emerging Markets Fund. EEM holds ADRs and GDRs in emerging market companies. EEM tracks the MSCI Emerging Markets Index.

EFA. The iShares MSCI EAFE Index Fund. EFA holds stocks from Europe, Australia, and the Far East. It tracks the MSCI EAFE Index.

EWA. The iShares MSCI Australia Index Fund. EWA holds stock in Australian companies. EWA tracks the MSCI Australia Index.

EWH. The iShares MSCI Hong Kong Index Fund. EWH holds stocks traded in the Hong Kong market and tracks the MSCI Hong Kong Index.

EWJ. The iShares MSCI Japan Index Fund. EWJ holds stocks traded on the Tokyo Stock Exchange. EWJ corresponds to the MSCI Japan Index.

EWK. The iShares MSCI Belgium Index Fund. EWK holds stock traded on the Brussels Stock Exchange. EWK tracks the MSCI Belgium Index.

EWP. The iShares MSCI Spain Index Fund. EWP holds stocks that are primarily traded on the Madrid Stock Exchange. EWP corresponds to the MSCI Spain Index.

EWS. The iShares MSCI Singapore Index Fund. EWS holds stocks traded on the Singapore Stock Exchange. EWS tracks the MSCI Singapore Index.

EWT. The iShares MSCI Taiwan Index Fund. EWT invests in stock traded on the Taiwan Stock Exchange. EWT corresponds to the MSCI Taiwan Index.

EWW. The iShares MSCI Mexico Index Fund. EWW holds stocks traded in the Mexican Stock Exchange. EWW tracks the MSCI

Mexico Index.

EWY. The iShares MSCI South Korea Index Fund EWY holds stocks traded on the South Korean Stock Exchange. It tracks the MSCI South Korea Index.

EWZ. The iShares MSCI Brazil Index Fund. EWZ holds stocks traded on the Brazilian stock market, the Bolsa de Valores de Sao Paulo. The performance of EWZ tracks the publicly traded securities in the Brazilian market.

EZU. The iShares MSCI EMU Index Fund. EZU invests in stocks from Austria, Belgium, Finland, France, Germany, Greece, Ireland, Italy, the Netherlands, Portugal, and Spain, and tracks the MSCI EMU Index.

FXI. The iShares FTSE/Xinhua China 25 Index Fund. FXI holds 25 of the largest and most liquid Chinese companies.

GLD. The streetTRACKS Gold Trust. Shares in GLD track the price of gold bullion.

IBB. The iShares Nasdaq Biotechnology Index Fund. IBB holds biotechnology companies. IBB tracks the Nasdaq Biotechnology Index.

ICF. The iShares Cohen & Steers Realty Majors Index Fund. ICF seeks investment results that track generally to the price and yield performance, before fees and expenses, of the Cohen & Steers Realty Majors Index.

IDU. The iShares Dow Jones U.S. Utilities Sector Index Fund seeks investment results that track the price and yield performance, before fees and expenses, of the Dow Jones U.S. Utilities Sector Index.

IEF. The iShares Lehman 7-10 Year Treasury Bond Fund IEF holds U.S. intermediate-term U.S. government Treasury bonds. IEF tracks the intermediate-term sector of the United States Treasury.

IGE. The iShares Goldman Sachs Natural Resources Index Fund. IGE holds stock in extractive industries, energy companies, owners of timber tracts, forestry services, producers of pulp and paper, and owners of plantations. IGE corresponds to the Goldman Sachs Natural Resources Sector Index.

IGM. The iShares Goldman Sachs Technology Index Fund. IGM holds stock in computer-related, electronics, networking, Internet services, and Internet software companies. IGM tracks the Goldman Sachs Technology Sector Index.

IGN. The iShares Goldman Sachs Networking Index Fund. IGN holds telecom equipment, data networking and wireless equipment producers. IGN tracks the Goldman Sachs Technology Industry Multimedia Networking Index.

IGV. The iShares Goldman Sachs Software Index Fund. IGV holds companies that produce client/server, enterprise, Internet software, PC, and entertainment software. IGV tracks the performance of the Goldman Sachs Technology Industry Software Index.

IGW. The iShares Goldman Sachs Semiconductor Index Fund. IGW holds manufacturers of wafers and producers of capital equipment for chip manufacturing. IGW tracks the Goldman Sachs Technology Industry Semiconductor Index.

ILF. The iShares S&P Latin America 40 Index Fund. ILF holds ADRs in companies from Mexico, Brazil, Argentina, and Chile. ILF tracks the S&P Latin America 40 Index.

IOO. The iShares S&P Global 100 Index Fund. IOO holds 100 common stocks from diverse countries. IOO tracks the S&P Global 100 Index.

IWM. The iShares Russell 2000 Index Fund. IWM holds the 2000 smallest capitalization-weighted companies in the Russell 3000 Index.

IYR. The iShares Dow Jones U.S. Real Estate Index Fund.

ICF holds stock in apartment and real estate investing companies. Tracks the Dow Jones U.S. Real Estate Index.

IYT. The iShares Dow Jones Transportation Average. IYT tracks the Dow Jones Transportation Index.

IYW. The Dow Jones US Technology Sector Index Fund. IYW invests in tech companies from software, telecommunications and semiconductor areas. IYW corresponds to the Dow Jones U.S. Technology Sector Index.

LQD. The iShares Goldman Sachs InvesTop Corporate Bond Fund. LQD holds investment-grade corporate bonds.

MDY. MidCap SPDR Trust. MDY holds mid-cap stocks and tracks the performance of the S&P stocks in the MidCap 400 Index.

~ Sector Trading: A Year in Exchange Traded Funds ~

ONEQ. Fidelity Nasdaq Composite Index Fund. ONEQ holds a portfolio of securities picked to correspond to the return of the Nasdaq stock market index.

QQQQ (QQQ before December 2004). The Nasdaq-100 Trust. QQQQ track Nasdaq-100 Index, which holds the largest stocks by market cap on the Nasdaq stock exchange.

RWR. SideTRACKS Wilshire REIT Fund. RWR invests in real estate. IYR tracks the Wilshire REIT Index.

SHY. The iShares Lehman 1-3 Year Treasury Bond Fund. SHY holds near-term U.S. government Treasury bonds.

SPY. Standard and Poor's Depositary Receipts (SPDRs) Trust. The SPY tracks the performance of the S&P 500 Index.

TIP. The Lehman TIPS Bond Fund. TIP holds U.S. Treasury inflation-protected notes. TIP tracks the Lehman Brothers U.S. Treasury Inflation Notes Index.

TLT. The iShares Lehman 20+ Year Treasury Bond Fund. TLT holds long-term U.S. government Treasury bonds and tracks the Lehman Brothers Bond Index.

XLB. The Materials Select Sector SPDR Fund. XLB holds chemicals, construction materials, containers, packaging materials, mining, and paper and forest products. XLB tracks the materials economic sector in the S&P Composite Index.

XLE. The Energy Select Sector SPDR. XLE holds companies in the oil, gas, energy equipment, and services sectors. XLE tracks the energy sector in the S&P Composite Index.

XLF. The Financial Select Sector Index. XLF holds stock in banks, insurance, and real estate. It tracks the financial sector in the S&P Composite Index.

XLP. The Consumer Staples Select Sector Index. XLP holds companies such as food and drug retailing, beverages, food products, tobacco, household products, and personal products. XLP tracks the consumer staples sector in the S&P Composite Index.

XLU. The Utilities Select Sector SPDR Fund. XLU has holdings in chemicals, construction materials, containers and packaging materials, mining, and paper products. It tracks the utility sector of the S&P Composite Index.

ETF Index

AGG. *6, 11, 13, 22-24, 37, 43, 48, 59, 61, 90, 100, 115, 123, 134, 174*

DIA. *6, 15-17, 20, 26, 30, 38, 50, 52-54, 76, 81, 89, 90, 96, 100, 105, 106, 108, 110, 113, 115, 122, 138, 149, 151, 154, 159, 162, 164, 168, 170, 171, 181, 186, 190-193, 198, 201-203, 206, 207, 210-212, 214, 215, 225, 240*

DVY. *146, 147*

EEM. *68*

EFA. *147, 148, 179, 183, 211, 212*

EMA. *5, 14-19, 31, 125, 126, 240*

EWA. *169, 172, 173*

EWH. *179, 183, 226*

EWJ. *59, 68, 82, 89, 93, 100, 109, 179, 183, 199, 204, 211, 212, 226, 238*

EWK. *109*

EWP. *3, 4*

EWS. *243*

EWT. *59, 110*

EWW. *82, 109, 141, 150, 161, 165, 183, 226*

EWY. *100*

EWZ. *6, 59, 93, 109, 141, 150, 161, 163, 165, 169, 172, 173, 183, 226*

EZU. *59, 82, 89, 199, 211, 212*

FXI. *226*

IBB. *6, 15, 18, 20, 23, 40, 61, 66, 75, 110, 123, 155, 157, 158, 178, 231*

ICF. *6, 11, 13, 19, 24, 78, 83, 86, 90, 109, 114, 117, 124, 127, 129, 130, 204*

IDU. *6, 7, 46, 47, 77, 86, 94, 98, 106, 119, 124, 134, 201, 224, 226, 228*

IEF. *6, 36, 43, 64, 71, 87, 88, 119, 129, 133, 134, 139, 140, 144, 155, 161, 169, 173, 184, 195, 196, 199, 200, 216*

IGE. *29, 39, 42, 45, 50, 57, 64, 67, 91, 93, 94, 118, 133, 135, 143, 162, 165, 189*

IGM. *6, 15, 18, 20, 23, 29, 33, 36, 37, 41, 58, 66, 102, 110, 114, 117, 119, 123, 182*

IGN. *15, 36, 38, 40, 65, 102, 114, 128, 129, 159, 166, 182, 199, 207*

IGV. *15, 29, 33, 58, 65, 97, 102, 114, 128, 159, 166, 182, 199, 207*

IGW. *6, 15, 20, 29, 33, 36, 46, 58, 60, 61, 64-66, 102, 110, 114, 124-126, 128, 129, 152, 159, 163, 166, 182, 184, 185, 189, 199, 207*

ILF. *109, 141, 150, 161*

IOO. *173*

IWM. *36, 63, 85, 96, 162, 178, 183, 210, 212*

IYR. *11, 58, 78, 90, 109, 116, 129, 184*

IYT. *53, 54*

LQD. *22, 23, 32, 36, 59, 87, 88, 90, 96, 100, 115, 119, 123, 191, 199, 200*

MDY. *36, 53, 63, 85, 96, 162, 178, 210, 212*

QQQQ. *15-17, 21, 26, 38, 50, 52, 55, 66, 70, 75, 76, 81, 88, 100*

RWR. *11, 58, 78, 83, 86, 90, 109, 114*

SHY. *43, 59, 64, 110, 123, 129, 139, 140, 144, 169, 173, 174, 184, 195, 199, 200, 216*

SPY. *6, 7, 11, 15, 17, 20, 25, 46, 52, 53, 58, 59, 67-69, 71, 75, 80, 82, 83, 85, 88, 100, 103, 108, 109, 113-116, 118, 122, 129, 138, 151, 153, 154, 159-161, 164, 166-168, 170, 171, 178, 179, 181, 186, 189, 190, 194, 198, 199, 202-204, 206, 207, 211, 212, 215, 219, 222, 223, 225, 228, 238, 240*

TLT. *6, 11, 22-24, 32, 36, 43, 59, 64, 65, 68, 69, 74, 79, 82, 83, 87, 88, 90, 91, 96, 101, 119, 123, 124, 129, 133, 134, 139, 140, 144, 155, 161, 169, 173, 174, 184, 191, 195, 196, 199, 200, 212, 216, 219, 225*

VIX. *187-189, 194, 196, 202, 203, 212, 213*

XLB. *246*

XLE. *6, 7, 20, 21, 29, 30, 37, 39, 41, 42, 44-46, 50, 55, 67, 69, 70, 77, 91-95, 98, 103, 109, 118, 124, 133, 135, 143, 150, 152, 157, 162, 165, 172, 174, 177, 189, 199, 203, 205, 208, 209, 212, 219, 222-224, 232*

XLF. *6, 15, 17, 105, 106, 115, 143*

XLP. *246*

XLU. *46, 77, 86, 94, 106, 119, 134, 212, 226*

Index

A

Apple Computer 125, 182
Asia 1, 39, 58, 182, 217, 226
Atkins Diet 74
Australia 100, 169, 173, 243
Austria 109, 244

B

Bear market 40, 162, 173, 187, 240
Belgium 109, 243, 244
Biotech 18-20, 27, 30, 33, 34, 40, 70, 75, 123, 155-159, 163, 179, 180, 184, 228, 235
Biotechnology 18-20, 27, 30, 33, 34, 40, 70, 75, 123, 155-159, 163, 179, 180, 184, 228, 235
Bonds
    Corporate bonds 87, 174, 200, 245
    TIPS Bond 11, 32, 191, 246
    Treasury bonds 11, 72, 139, 145, 161, 191, 243, 244, 246
Boredom 48
Brazil 6, 59, 93, 109, 141, 150, 151, 161, 163, 165, 166, 169, 173, 183, 226, 227, 244, 245
Broad market ETF 18, 26, 85, 108, 130, 132, 188, 192, 240
Bull market 5, 43, 52, 57, 65, 134, 162, 236, 240
Bush, George W. 5, 30, 34, 39, 40, 47, 79, 81, 82, 84-86, 97, 110, 122, 226, 237

C

Canada 109
Casino 48
Cheney, Dick 30
Chinese 39, 136, 160, 196, 218, 226, 238, 244
Congress 142, 217, 237
Construction spending 58
Consumer confidence 28, 109
Consumer Price Index 22, 122, 145, 166, 202, 240
Consumer Price Index (CPI) 22, 145, 240
Corruption
    Enron iv, 31, 235
    Tyco iv
    WorldCom iv
Cubes 75, 83, 103, 166, 185, 196, 198, 201, 208
Currency
    Euro 1, 2, 58, 59, 68, 82, 89, 93, 96, 100, 116, 119, 144, 179, 183, 199, 204, 211, 238, 243
    Pound 156
    Yen 82, 93, 96, 100, 199

D

Deflation 144, 145
Dividends 46, 72, 78, 79, 116, 120, 146, 228
DOT limit iv, 240
Dow Jones Industrial Average 15, 26, 38, 50, 53, 76, 227, 243
Drugs 156
Durable goods 28, 55, 93, 128, 142, 190, 206
Duration 23, 24, 59, 199, 240

E

Election 3, 5, 8, 76, 79, 81-83, 85, 87, 89, 105, 122, 124, 129, 132
Employment 5, 32, 63, 65, 96, 114, 132, 133, 154, 155, 210
Energy
    Energy sector 14, 15, 29, 30, 39, 55, 67, 91, 133, 152, 199, 203, 246
    Index 223
Equity market 90, 109, 124, 161
Etfzone 2
Etfzone.com 2, 5
Europe 1, 2, 58, 59, 68, 82, 89, 93, 119, 179, 183, 199, 204, 211, 238, 243
European Central Bank 119
Existing home sales 28, 206

F

Fat cats 44
Fear 6, 12, 13, 32, 36, 47, 49, 87, 114, 116, 138, 139, 144, 147, 149, 150, 152, 153, 179, 186, 187, 194, 196, 202, 213, 217, 225
Fed model 5, 71-75, 79, 83, 87, 91
Federal Open Market

Committee 241
Federal Reserve 8, 10, 11, 25, 65, 71, 72, 87, 116, 124, 132, 140, 168, 217, 235, 237, 238, 241
Ford Motor 8, 30, 34, 52, 194, 205
France ii, 195, 244
Futures 30, 41, 45, 149, 164, 171, 172, 177, 189

G

Gambling 48
GDP 28, 76, 128, 129, 174, 190, 207
General Electric 2, 3
General Motors 6, 15, 18, 20, 23, 29, 33, 36, 37, 41, 58, 66, 102, 110, 114, 117, 119, 123, 133, 182, 194, 200, 244, 245, 247
Gold
    GLD 115, 146, 148, 160, 244
Greed 6, 37, 99, 136, 149, 150, 152
Greenspan, Alan 12, 25, 90, 91, 117, 142, 143, 168, 169, 203, 204, 215, 217, 235, 237

H

Haseltine, William
    Haseltine 156, 157
Health care 15, 18, 104, 105, 110, 178
Housing starts 24, 122, 202

I

Inflation 6, 22, 32, 34, 49, 89, 90, 99, 100, 113, 118, 119, 124, 132, 141, 142, 144-148, 160, 168-170, 191, 196, 202, 207, 214, 216, 217, 240, 242, 246

Interest rate 11-13, 25, 32, 68, 74, 76, 87, 117, 120, 132, 136, 144, 146, 147, 169, 183, 237, 240-242
Investor confidence 187
Iraq 132, 136

J

Japan 59, 68, 82, 89, 93, 100, 109, 144, 145, 179, 183, 199, 200, 204, 211, 226, 238, 243

K

Kerry, John 79, 82, 84
Korea ii, 100, 244

L

Labor Department 10, 32
Latin America 1-3, 58, 109, 130, 141, 150, 161, 165, 226, 245
Livermore, Jesse v, 162
Long maturity 88
Love affair 200

M

Manufacturing 190, 245
Market Factors
    Bond Yields 2, 8, 10, 21, 32, 85, 122, 132, 138, 142, 149, 160, 168, 171, 194
    Currency 8, 57, 67, 81, 89, 93, 95, 99, 108, 113, 118, 142, 160, 168, 177, 198, 202, 210, 215, 225
    Energy 8, 21, 28, 38, 42, 45, 49, 55, 57, 63, 67, 71, 76, 81, 89, 93, 95, 99, 102, 108, 118, 132,

138, 149, 160, 164, 171, 177, 181, 186, 190, 194, 198, 202, 206, 215, 221, 225
    Fundamental Analysis 8, 28, 35, 71, 76, 104, 108, 122, 149, 154, 225
    Government economic data 8, 10, 21, 28, 32, 35, 42, 45, 49, 52, 55, 57, 63, 76, 81, 89, 95, 102, 113, 118, 122, 128, 132, 138, 142, 154, 160, 181, 190, 198, 202, 206, 210, 221
    Investor psychology 8, 14, 25, 32, 35, 38, 67, 76, 149, 160, 181, 186
    Market rhythm and counting 8, 25, 38, 63, 67, 95, 206
    Political decisions 8, 42, 45, 49, 81, 85, 122
    Seasonal factors 8, 10, 25, 35, 45, 49, 57, 93, 99, 142, 181, 190, 206, 215, 221
    Technical analysis 8, 14, 21, 25, 32, 35, 38, 52, 55, 67, 71, 85, 104, 108, 118, 122, 149, 186, 190, 215
Market Rally 5, 6, 14, 34, 38-40, 46, 51, 54, 60, 74, 76, 85, 99, 105, 110, 116, 122, 124, 137, 174, 180, 186, 187, 191, 195, 202, 203, 207, 208, 212, 228, 234, 244
Materials Index 57
Mexico 53, 82, 95, 109, 141, 150, 151, 161, 165, 166, 183, 226, 227, 243-245
Mid-cap 63, 85, 86, 96,

132, 161, 162, 178, 179, 183, 210-212, 241, 245

Money managers 47, 61

Mutual funds 1, 2, 241

N

Nasdaq i, 6, 7, 14, 15, 18, 21, 22, 26, 33, 34, 86, 110, 122, 157, 184, 206, 207, 231, 244, 246

Nassim, Taleb 69

Netherlands 244

Niederhoffer, Victor 39, 40

O

Oil
    Futures 41, 45, 149, 164, 172, 177, 189
    Gasoline 28, 30, 135, 136, 171
    OPEC 6, 99, 132-137, 150, 165, 177, 221, 241
    Service Sector 14, 30, 37, 44, 69, 92, 135, 172, 177, 180, 236

ONEQ 26, 34, 246

P

Politics 5, 8, 20, 44, 45, 49, 122, 136, 235

Producer Price Index 6, 36, 39, 49, 89, 90, 99, 118, 142-146, 166, 227, 241, 242

Producer Price Index (PPI) 6, 49, 89, 99, 118, 142, 241

Profit-taking 116, 178

Psychology 8, 185

Q

Qualcom 208

R

Random 5, 67-70

Real Estate
    General 1, 2, 11, 109, 110, 117, 144, 218, 245, 246
    REITs 5, 10-13, 20, 24, 27, 58, 59, 78, 79, 83, 86, 90, 114-117, 120, 124, 129, 130, 178, 184, 204, 205, 236

Republican National Convention 44, 45, 49

Retail sales 68, 164, 182, 198, 199

Rice, Condoleezza 30

Risk 3, 4, 24, 30, 37, 44, 47, 59, 72, 78, 79, 83, 87, 91, 104, 120, 126, 136, 147, 148, 169, 172, 184, 185, 205, 207, 227, 234, 236, 237, 240-242

Rothschild, Nathan 189

S

S&P 500 15, 38, 72, 73, 75, 86, 166, 195, 203, 246

Saudi Arabia
    Country 136, 165, 209, 222
    King Fahd 209

Shorting stock 31

Sideways market 5, 41, 99, 101, 173, 192

Soros, George 5

South Africa 109, 170, 173

Spain 3, 4, 243, 244

Stewart, Martha 74

T

Taiwan 59, 110, 243

Taleb, Nassim 69

Target federal funds rate 117

Taxes 30, 185

Technology
    DRAM chips 208
    Index 6, 15, 18, 244
    Sector 5, 15, 18, 19, 21, 23, 36, 46, 51, 61, 64, 65, 110, 124, 182, 207, 245
    Software 5, 15, 29, 33, 58, 65, 97, 102, 114, 125, 128, 159, 166, 182, 199, 207, 208, 215, 245

Telecommunications 245

Terrorism 12, 44, 47, 74, 76, 83

Thomas Weisel Partners 72

TIP 11, 32, 100, 145-147, 191, 246

Top-performing 126

Trade deficit 35, 68, 138, 160, 182, 198, 199

Transportation 5, 37, 52, 53, 245

Treasury 6, 11, 71-74, 90, 110, 133, 139, 140, 145, 161, 191, 195, 199, 214-219, 237, 238, 240, 243, 244, 246

Tysabri 155, 158

U

U.S. Energy Information Administration 172

Utilities 5-7, 46, 47, 76-79, 84, 86, 94, 97, 98, 101, 103, 106, 110, 115, 119, 120, 124, 134, 137, 170, 178, 180, 185, 193, 223-226, 228, 235, 244, 246

V

Venezuela 136
Volatility 4, 6, 23, 29, 43, 46, 51, 79, 88, 90, 97, 100, 126, 129, 130, 135, 141, 152, 162, 163, 171, 183, 186-189, 191, 195, 202, 203, 210, 212, 213, 219, 223, 228, 231, 234, 240, 242

Volume 8, 10, 27, 34, 37, 45, 152, 179, 181, 211, 215, 238, 242

W

Wholesale prices 142-145
Wilderhill Index 223

Printed in the United States
105973LV00003B/113/A